ANALYZING SAFETY SYSTEM EFFECTIVENESS

ANALYZING SAFETY SYSTEM EFFECTIVENESS

Third Edition

Dan Petersen

JOHN WILEY & SONS, INC.
New York • Chichester • Weinheim • Brisbane • Singapore • Toronto

A NOTE TO THE READER
This book has been electronically reproduced from digital information stored at John Wiley & Sons, Inc. We are pleased that the use of this new technology will enable us to keep works of enduring scholarly value in print as long as there is reasonable demand for them. The content of this book is identical to previous printings.

Copyright © 1996 by John Wiley & Sons, Inc. All rights reserved.

Originally published as ISBN 0-442-02180-1

No part of this publication may be reproduced, stored in a retrieval system, or transmitted in any form or by any means, electronic, mechanical, photocopying, recording, scanning or otherwise, except as permitted under Sections 107 and 108 of the 1976 United States Copyright Act, without either the prior written permission of the Publisher, or authorization through payment of the appropriate per-copy fee to the Copyright Clearance Center, 222 Rosewood Drive, Danvers, MA 01923, (978) 750-8400, fax (978) 750-4744. Requests to the Publisher for permission should be addressed to the Permissions Department, John Wiley & Sons, Inc., 605 Third Avenue, New York, NY 10158-0012. (212) 850-6011, fax (212) 850-6008, E-mail PERMREQ@WILEY.COM.

For ordering and customer service, call 1-800-CALL-WILEY.

Library of Congress Cataloging-in-Publication Data:

Petersen, Daniel
  Analyzing safety system effectiveness / by Daniel Petersen. —3rd ed.
    p. cm.
  Revised ed. of: Analyzing safety performance.
  Includes bibliographical references and index.
  ISBN 0-471-28739-3 (hardcover)
    1. Industrial safety. 2. Industrial safety—Management.
I. Petersen, Daniel. Analyzing safety performance. II. Title
T55.P345   1996
658.4'08—dc20   96-1693
                                                              CIP

Contents

Preface to the Third Edition   ix
Preface to the First Edition   xi

PART I.   APPROACHES TO ANALYZING PERFORMANCE

Chapter 1. The Analysis Task   3
   What Must Be Done: The Three-Step Process   3
   The Safety Professional's Role   4
   Safety Management Principles   7
   Management's Role   10
   The Employee Role   10
   Culture Is the Key   11

Chapter 2. How We Traditionally Have Analyzed   15
   System Approaches   16
   Areas to Analyze   25
   Benchmarking   29

Chapter 3. Analysis by Workers   33
   The Perception Survey   33
   Interviewing   36
   Safety Behavior Sampling   40

PART II.   THE AREAS TO ANALYZE

Chapter 4. The Management System for Continuous Improvement 47
  Your Accident Investigation Procedures 47
  Employee Involvement and Participation 55
  Operating Procedures and Practices 57
  Discipline 59

Chapter 5. The Management System to Build Culture 65
  Safety Climate (or Culture) 65
  Management Credibility 68
  Support for Safety 71
  Recognition for Performance 74
  Attitudes toward Safety 75
  Stress 81

Chapter 6. The Management System to Improve Managers' Skills 89
  Supervisory Training 89
  Quality of Supervision 92
  Goals for Safety Performance 95

Chapter 7. The Management System to Improve Employee Skills 101
  Employee Training 101
  New Employees 105
  Communications 108

Chapter 8. The Management System to Improve Worker Behavior 115
  Safety Contacts 115
  Alcohol and Drug Abuse 117
  Awareness Programs 121

Chapter 9. The Management System to Improve Physical Conditions 127
  Inspections 127
  Hazard Correction 130

Chapter 10. Additional Areas to Analyze 135
  In the Management System for Continuous Improvement 136
  In the Management System to Improve Behavior 150
  In the Management System to Improve Physical Conditions 158
  In the Management System to Contain Costs 166

PART III. THE CHANGE PROCESS

Chapter 11. Interpreting the Data 173
  Steps to Survey Results Interpretation 176
  Analysis by Unit 184

Chapter 12. Where Do You Want to Be? 185

Contents

>*The Management System and Accidents 187*
> *The Function of Safety 187*
> *The Ten Basic Principles of Safety 188*

Chapter 13. Defining and Implementing Change 193
> *Fixing Problems: An Example 195*
> *Involvement Approaches 198*

Chapter 14. Sources of Help 209
> *When to Seek Outside Help 209*
> *What External Help Is Available 210*
> *Choosing a Consultant 223*
> *Analyzing a Consultant's Performance 223*
> *Internal Problem Solving 228*

PART IV. APPENDIXES

> *Appendix A. Measurement 233*
> *Appendix B. Safety Sampling 239*
> *Appendix C. Accountability Systems 247*
> *Appendix D. Supervisory Tasks 253*

Bibliography 263

Index 267

Preface to the Third Edition

The first edition of this book was published in 1980, so the information was relevant through about 1977 (publication takes a while). The second edition was published in 1984, and since there were few changes, it too was relevant through about 1977. At the time of this writing it is 1995, nearly two decades later.

What has changed since 1977 besides the author's getting older? Just about everything in the field of safety! This book was far enough out of date several years ago that we decided to retire it—to remove it from publication; and we did. We considered the information to be no longer relevant, to be out of date. For we had written that there were essential elements to a safety program, that doing certain things would get results. We thought that to be true in 1977. Much has changed; we know better today—that there are *no* essential elements, that what works in another company may not work in yours, that it is not the elements that count but rather the climate in which you place those elements that determines their effectiveness. Safety meetings, or JSAs or whatever, can be highly effective in Company A and a total waste of time in Company B (or even counterproductive).

Although this is clear to anyone who has spent any time in industry, it apparently is not obvious to some others. In the United States, OSHA published the "Guidelines to a Safety Program," which have become law in California (SB 198), Oregon, and other places. Other guidelines are being prepared (such as those for ergonomics). Therefore, the painfully clear message from research, from world class companies, and from those companies that have made even step-change improvement obviously has not been received.

My hope in publishing this new edition is simply to say once again that even if government regulates, you and I can still get something accomplished in our organizations by building a system based on what we know works from research, benchmarking, and good managerial and behavioral concepts, rather than just following regulatory dictates.

Preface to the First Edition

As a safety professional, I have long been impressed with the earlier work of those professionals in the American Society of Safety Engineers who originally, in the 1960s, came up with the document "Scope and Functions of the Safety Professional." The document stated rather precisely that the functions of the safety professional are fourfold; to analyze, to develop plans, to communicate those plans to the line, and to monitor.

Function one of the four—to analyze—is by far the most important; for all progress is based on the proper analysis of the situation that exists. If we misdiagnose, our plan of attack will be misdirected. It is only through proper diagnosis that we can improve the record of our organization. In short it is extremely important.

This book describes three areas of analysis: the physical, the managerial, and the behavioral. All are important, and all must be included if our analysis is to be valid. Usually we have concentrated on only one area; or perhaps two if we have been sophisticated in our approach. Invariably, however, we have not hit all three.

This book does not attempt to prescribe solutions to all safety problems; we have had far too much of that in traditional safety approaches. Rather, this book attempts to urge a better approach to analysis; even offer some different approaches you may wish to consider.

In some sections of this book I have quoted several well known authors rather heavily. What they have to say needed to be included in their own words. To them I am deeply grateful. Their names will be obvious to you as you read through this book.

PART I | Approaches to Analyzing Performance

Chapter **1**

The Analysis Task

WHAT MUST BE DONE: THE THREE-STEP PROCESS

Any organization that chooses to control losses can do so. The process it must follow is exceedingly easy, consisting of three small steps:

1. Deciding where it wants to be.
2. Determining where it is now.
3. Providing the difference.

This is the management and/or training model that we have been using for years. Management uses it every day, almost innately. Training people should use it every day (but usually do not). Safety people seldom use it.

This book is about how we can perform this process: how we can assess accurately and validly where we are in our safety program, and how we can decide where we want to be. My experience is that discerning the difference and how to achieve it will follow naturally.

There is one major problem in the process. In order to determine where our safety system should be, and in order to accurately assess where our system is right now, we may have to consider some alternative ideas to what we know; to what we have always believed about safety; about what works and what does not.

If you are not ready to do that, then it would probably be better for you to take this book back to wherever you got it, and forget the whole thing. I cannot conceive of any organization getting a handle on its safety problems today without turning in its old ideas and beliefs first.

What are the old beliefs that need to be turned in? Here are a few for starters:

- That accidents are caused by unsafe acts and conditions.
- That there are certain "essential elements" to a safety program.
- That accident statistics tell us anything.
- That audits predict results in safety.
- That regulatory compliance ensures safety results.

Thus the three-step process is easy:

1. What do you think your safety system should look like? What should it consist of?
2. What is your system now? What does it look like? What does it consist of?
3. Determine an action plan to get from point A to point B.

We first consider who does the three-step process.

THE SAFETY PROFESSIONAL'S ROLE

In 1963 the executive committee of the American Society of Safety Engineers initiated a plan designed to identify the type of work that safety personnel should be doing. This was part 1 of a three-part major project; it described the scope and functions of the safety professional's position. Part 3 developed a curriculum or formal course of study, leading to a university degree, which would prepare the safety professional to perform the functions described in part 1. Part 2 established procedures leading to the acquisition of some form of certification or registration as a means of demonstrating competence in the field.

Scope and Functions

The major functions related to the protection of people, property, and the environment are:

- Anticipate, identify, and evaluate hazardous conditions and practices.
- Develop hazard control designs, methods, procedures, and programs.
- Implement, administer, and advise others on hazard controls and hazard control programs.
- Measure, audit, and evaluate the effectiveness of hazard controls and hazard control programs. (Figure 1-1).

The Analysis Task

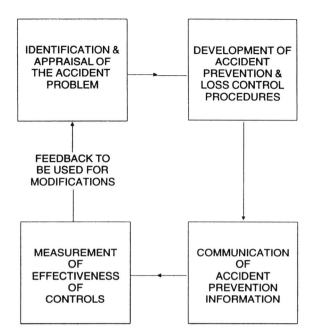

Figure 1-1. The functions of the safety professional. (Reprinted by permission from ASSE, paper entitled "The Scope and Functions of the Safety Professional," undated.)

The original scope and functions document is reflected in the most recent publication of the ASSE, although now slight changes from that document indicate the direction of the profession. There have been no changes in the four primary functions, but there are differences in how some of them are defined.

For instance, in all earlier documents the first role was to identify and appraise the management system to determine what system failures were present that could result in loss, in the physical, managerial, and behavioral environment. This role has now been expanded to include anticipation, identification, and evaluation of the changes.

In the earlier documents, the safety professional's role was to communicate to the line management what must be done to control loss, recognizing that the only persons who could control losses were in the line. In the most recent document, the role has become to implement, administer, and advise, removing the safety professional from a purely advisory, change-agent role to one of an active participant, allowing line managers the opportunity to abdicate their previously well-defined role.

Also, the safety professional's role used to be to measure and evaluate progress, allowing a choice of which measures would do the most valid job of measuring safety system effectiveness. In the recent document, audit is defined as the measure (or one of the measures) to be used. This presents no problem if the audit used correlates with the accident record over time (auditing is, in fact, what will get results); but if the audit is checking regulatory compliance primarily, it could be counterproductive, as is discussed in this book.

The safety professional's functions, then, are to appraise what exists, develop improvement plans, communicate those plans to the line organization, and measure results. These four areas refer to the safety aspects of the safety professional's job. He or she also may have many other safety-related (and non-safety-related) parts to the job. A second major part of the job of most safety professionals is that of ensuring that the organization is ready for OSHA. Today a third part is cost containment.

OSHA Responsibilities

After defining the safety responsibilities of the job, the safety director must go back and define the OSHA responsibilities. This process is not often similar to that of defining safety tasks. The safety director's role in OSHA compliance is occasionally self-defined, but usually it is not. In terms of OSHA compliance, the safety director assumes a quite different role, that of the corporate OSHA expert—the only one who "knows" the standards. Probably the reason for this is that the standards are written in such a way that the line organization perceives them as being much more technical and difficult to understand than they really are. As a result, the safety specialist is not able to ask line managers to ensure that their departments are in compliance.

Obviously, it did not have to be this way. There is no reason why a line manager could not ensure compliance if the relatively simple concepts covered by the standards had been simply and understandably written. In any event, the safety director seems to be stuck with a role in OSHA compliance that can usually be described as follows:

1. The safety specialist must interpret the physical standards for all in the company. Most line managers are simply not willing to expend the time and energy necessary to interpret "federalese."
2. The safety specialist must sift through those pages to find the various administrative requirements tucked in there (there were nearly 400 such requirements at last count). Personal experience tells me that no line manager would do this.
3. The safety specialist then must find the physical and administrative violations existing in the organization.
4. The safety specialist must assign priorities to the violations found and begin to schedule corrections.
5. The safety specialist must ensure that the corrections are accomplished.

The Analysis Task 7

6. The safety specialist then must construct a plan to assure that the organization will stay in compliance with the law.
7. Finally, the safety specialist must document everything that has been done to achieve compliance.

Safety Responsibilities

This book discusses only the professional's safety responsibilities and does not attempt to look at his or her nonsafety functions or safety-related functions. This is not to say that these allied functions are not important, because of course they are; OSHA compliance, for instance, is a legal responsibility of the organization for which the safety professional works. Usually that organization depends upon the safety professional to ensure compliance with the standards promulgated. These standards, and thus this function, are sometimes an integral part of the safety function. Sometimes the functions are very closely related; at other times the relation between OSHA compliance and achievement of safety is not so clear. There are many instances and many OSHA standards that are not tightly related to safety. In some instances complying with the standards is counterproductive to safety.

Our purpose is to concentrate solely on the safety responsibilities as designated in the ASSE scope and functions document, and to concentrate on only two of those functions. This book is about identification and appraisal of accident problems, and about the development of accident prevention and loss control procedures to deal with the identified problems.

SAFETY MANAGEMENT PRINCIPLES

Before discussing the basic approaches to analysis and development of accident-prevention plans and programs, we will examine briefly some of the underlying principles of these two functions. In 1971, in the first edition of my book *Techniques of Safety Management*, a number of the principles were discussed in some detail. These principles, from various sources, concerned such things as accident causation, the relationship between frequency and severity, the management of safety, the importance of accountability and measurement, and so on. The fifth principle in that list is the rationale for this entire book:

> The function of safety is to locate and define the operational errors that allow accidents to occur. This function can be carried out in two ways: (1)

by asking why accidents happen—searching for their root causes—and (2) by asking whether certain known effective controls are being utilized.

The first part of this principle was taken from the ideas of W. C. Pope and Thomas J. Cresswell as put forth in their article "Safety Programs Management," in the August 1965 issue of the *Journal of the American Society of Safety Engineers.* This article described safety's function as locating and defining operational errors involving (1) incomplete decision making, (2) faulty judgments, (3) administrative miscalculations, and (4) just plain poor management practices.

Pope and Cresswell suggested that to accomplish our purposes, we in safety would do well to search out not what is wrong with people but what is wrong with the management system that allows accidents to occur.

In further discussing the four major areas under identification and in appraising the problem, Pope and Cresswell stated that it is the function of safety to:

> Review the entire system in detail to define likely modes of failure, including human error, and their effects on the safety of the system. . . .
> Identify errors involving incomplete decision making, faulty judgment, administrative miscalculation and poor practices. Designate potential weaknesses found in existing policies, directives, objectives, or practices.

This concept directs the safety professional to look at the management system, not at acts and conditions.

The second part of the principle suggests that a two-pronged attack is open to us: (1) tracing the symptom (the act, the condition, the accident) back to see why it was allowed to occur, and (2) looking at the company's system (procedures) and asking whether certain things are being done in a predetermined manner that is known to be successful.

This approach has been well described by D. A. Weaver, as shown in Figure 1-2. In discussing the accident process, Weaver looked at the old "domino theory" of accident causation, originally from the writings of H. W. Heinrich, and suggested as an adaptation of the theory that the unsafe act and condition be viewed not as causes of accidents but rather as symptoms of things wrong in the management system. This ties in with the principle of multiple causation, as expressed by this principle from *Techniques of Safety Management:*

> An unsafe act, an unsafe condition, and an accident are all symptoms of something wrong in the management system.

We know that many factors contribute to any accident. Our thinking, however, has always suggested that we select one of these as the "proximate" cause of the accident or that we select one unsafe act and/or one unsafe condition. Then we remove that condition or act.

The Analysis Task

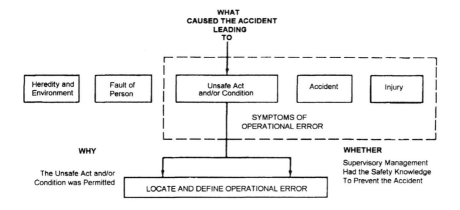

Figure 1-2. Weaver model of the accident process. (From D. A. Weaver, Symptoms of operational error, *Professional Safety,* October 1971.)

The theory of multiple causation suggests, however, that we trace all the contributing factors to determine their underlying causes. For instance, the amputation of a finger in a power press might start with an act (putting the hand under the die) and a condition (an unguarded point of operation). Tracing back from this point might lead, however, to an inquiry concerning why the operator was selected for the job, why the operator was poorly trained, why the supervisor allowed the act, why the supervisor was poorly selected and trained, why the maintenance on the press was poor, why the policy of management allowed an unguarded press, and so on.

This principle suggests not that we boil down our findings to a single factor but rather than we widen our findings to include as many factors as seem applicable. Hence every accident opens a window through which we can observe the system, the procedures, and so forth. Different accidents would unearth similar things that might be wrong in the same management system. Also, the theory suggests that, besides accidents, other kinds of operational problems result from the same causes. Production tie-ups, problems in quality control, excessive costs, customer complaints, product failures, and so on, have the same causes as accidents. Eliminating the causes of one organizational problem will eliminate the causes of others.

If we were to utilize this theory, we would redesign our accident-investigation procedures in a way that would enable us to identify as many contributing factors as possible in any single incident. Most of the changes would be directed at improving the organizational system—not at finding fault.

Identifying the act, the condition, and the accident only as a starting point to learn why the act and the accident were allowed to happen and why the condition was permitted to exist will lead to effective loss control. We view the accident, the act, and the condition as symptoms of something wrong in the system. Then we try to identify what is wrong in the organizational system that allows an unsafe act to be performed and an unsafe condition to exist.

If in the instance of the finger amputation we said merely that the cause was the unguarded press, the correction would consist in putting a guard on the press. However, we would have treated only the symptom, not the cause, because tomorrow the press would again be unguarded. This principle applies to any accident. Only when we diagnose causes and treat them do we effect permanent control. The function of the safety professional, then, is similar to that of a physician who diagnoses symptoms to determine causes and then treats those causes or suggests appropriate treatment.

This book is based upon these approaches to safety management. The focus will be on the first two functions of the safety professional—appraisal and development—and the principles involved are the search for "whys" and "whethers" in management systems.

MANAGEMENT'S ROLE

Management's role used to be easy—there was none. Top management used to sign policy, and first line supervisors were the "key people," but they really did not have to do anything except be lucky. This is not true today; the world has changed.

What is the management role today? The manager must play a very active role to keep from willful violations, to keep from going to jail, to keep the company above water financially, and so on.

Therefore, it seems obvious that the line manager, from CEO to first line supervisor, also needs to play a very active role in the three-step process discussed in this book. How this role works out in terms of specific activities depends entirely on the organizational situation, but the role today *must* be an active one. I have seen CEOs jump right in solving problems. First line supervisors in most companies spend considerable time providing their input to corporate solutions of safety problems.

Similar to this role, and of crucial importance, are the input and the involvement of the individual worker.

THE EMPLOYEE ROLE

The role of the employee is the most important one of all in the analysis process. As that person is the *only* person in the organization who really knows what is going wrong—what is successful and what is not—his or her input is absolutely crucial to determining the effectiveness of the organization's safety system.

CULTURE IS THE KEY

The elements of a safety program—such as safety policies, manuals, meetings, committees, inspection, investigation of incidents, and recordkeeping and analysis of records—are tools of the trade in most companies.

Different organizations use these same tools, but some have more success than others. How can we explain the variance?

Imagine two different safety scenarios, where two organizations faced the same situation. An accident occurred. Both organizations investigated as a part of their program.

The first organization had a clear policy to follow up on all incidents, and the investigation was launched. The investigation reached the supervisor, who had discovered that a worker did not adhere to the safety policy and narrowly escaped serious injury to himself and others. The employee was told of his unsafe behavior, and the safety infraction was noted. The employee promised not to engage in the unsafe practice again, and the necessary safety records were updated. Management recognized this supervisor for fostering workplace safety.

In the second organization's investigation, the supervisor considered the circumstances of the accident, namely, that it occurred when production pressures were greatest. The supervisor discovered that the worker was ill while under severe pressure to meet production deadlines. The supervisor also acknowledged that mechanical problems had slowed production that week, increasing frustration among workers and management. Company officials also acknowledged that maintenance must be done immediately, and that the company must reinforce its safety commitment by allowing the time to perform necessary maintenance between each job.

The second organization's investigation also revealed that recent company cutbacks had all employees concerned about job security—drawing attention away from safety practices and documentation. Workers petitioned upper management for the necessary actions and changes. Managers planned a meeting with all employees to discuss the current financial and organizational issues. They asked workers to maintain safety while working together to improve production because those things would help the corporation's viability.

What is the difference between these organizations? Why did one company blame the employee, fill out the incident investigation forms, and get back to work while the other company found that it must deal with fault at all levels of the organization?

The answer is "culture." A group of assembly line workers in a plant defined culture several years ago. Culture is "the way it is around here." A company's culture is at the heart of how safety system's elements or tools—such as incident investigations—are used. The culture is also expressed in the way company values are demonstrated. What happens when the pressure is on to get a job out the door and workers seemingly must choose between doing the job safely or doing it quickly?

The culture clearly announces every day to every worker whether safety is a key value and where it fits into the priorities. It dictates how employees will act and how they will be treated. As a result, it also dictates behavior (hard work, goofing off, or working safely). We know, to a large degree, that the organization's culture will determine the extent of casualties, trauma disorders, stress claims, and compensation paid. It dictates whether elements of a safety system will work or flop.

As the culture of the organization is so important, and the only people who truly know the culture are the hourly workers, obviously these are the people who must be asked to assess safety system effectiveness.

It took me many years to understand the importance of culture in determining success with safety systems. Like most safety professionals, I used to believe certain elements were essential to a safety program. I preached that all companies should include them—while I continued to see that safety meetings were successful in one company and a total joke in the next. Job safety analysis (JSA) was the fundamental thrust of a safety program at one company and a total waste of time at the other.

Safety programs work within cultures where people use the elements to support their safety goals. The elements are not the safety programs. The elements are the tools that managers, supervisors, and workers use. They can even provide structure for employees to discover where problems in the organization exist. Achieving alignment or agreement among employees on how the safety elements will be used is essential to create a true safety culture. This approach is participative, positive, and flexible; it has upper management support, supervisors' accountability, and middle managers actively involved.

Safety Culture Obstacles

One obstacle is a perceived lack of management commitment. Safety professionals have often complained that they must continually "plead for management support" for safety programs.

Lack of management commitment is more a perception than reality. I have seldom found an executive who was not genuinely interested in worker safety. Executives may be frustrated because they do not understand how to "fix" their safety program, but there is rarely a true lack of management commitment.

The Analysis Task

Top management generally wants to build a new organizational culture (even build safety into that culture), and workers genuinely want a changed, modern, and different safety effort. However, it does not always get done.

Executives with vision want a change in the culture. They want safety to be a key value. However, there are often barriers between the CEO and the worker. These barriers are created by the middle managers, who may resist change. Safety directors often find themselves a part of this group, which can add to their own frustration.

Typically, the three echelons each work together (although they do not know it) to sustain old cultures. A positive change requires someone first to recognize the nature of the problem and make the current culture visible.

Here is how the stalemate works:

Top managers tend to feel frustrated because they clearly state the safety value, trying to sell their vision, but no one seems to step forward and take charge. Perhaps the safety problem has not been clearly defined because it is shrouded in the culture and cannot yet be solved. It often appears that programs and systems should work, but the desired results do not emerge.

Middle managers tend to feel powerless. Safety directors know this feeling. They hear (what may be) "rhetoric" from upper management but are confronted with the reality of mixed messages caused by such things as cost cutting and downsizing. Safety gets lip service, but meeting production schedules is the priority. It is easy for them to be cynical because they have been in the organization long enough—longer than the sometimes relatively youthful, more recently educated upper managers—to see various programs come and go with little lasting impact (those programs did not fit or survive the culture either). Middle management tends to be less well trained in some of the newer management philosophies than upper management. Since middle managers are usually held responsible for production, they will typically opt to get the product out rather than hold the line on other values.

Workers know there are problems. They can even tell you where many of the problems are, but they do not have an overall solution. They tend to feel undervalued because no one asks them their opinion. Workers observe the lip service of top management but experience mixed messages when pressure and stress spell out clearer priorities: "Just get the work out." Feeling at risk, knowing it could be better, but resigned about the current situation, they have little left to do but complain.

As long as nothing happens to break this stalemate, the old culture will persist. Someone in any one of the employee work groups can demonstrate leadership and create the pathway for change. The opportunity is for those responsible for the safety programs to recognize what is going on and take positive action. What must be done is to reveal the culture, the atmosphere, that enables the safety program to occur.

The Perception Survey

The most powerful and effective thing to do is to conduct a perception survey to determine what is really happening. The idea is to ask people in the organization, anonymously, what they think. This can be done using a statistically valid survey tool and/or an interview process.

Once upper management has a clear picture, swift and decisive action is taken. For example, many interviews were conducted recently in a survey done for a utility company, and a pilot perception survey was done on one of its units. About 500 people were interviewed in a month, and then the CEO was presented the results. "I was afraid it might be this bad, and you confirmed my worst suspicions," he said. He described getting the findings as like an out-of-body experience, and he took immediate action to begin correcting the situation.

Top executives frequently operate with the illusion that they are making some headway, and that things are not as bad as they really are. This attitude can be a significant obstacle to the change process.

The perception survey makes it possible to ask the workers what they see and how they feel in a comprehensive and complete way. It may be the first time they think that they can be really truthful. Basically, the survey provides a pipeline of information directly from the workers to upper management, bypassing the blockage of middle management. After you have received the information, the change process begins, and it must involve people from all levels in the organization to avoid resistance.

The key is involvement. Creating a new safety culture is a continuous process that never really ends. All of the issues identified by the perception survey, or through interviews, cannot be solved at once. The approach is to form problem-solving work groups representing a cross section of the organization. Work groups made up of managers, supervisors, and employees are formed to study each of the lowest-rated safety categories, to analyze the problems, and to devise a plan that resolves them.

Chapter 2

How We Traditionally Have Analyzed

You probably want to be world class—the best in your industry. To quote a few of the companies I have worked with: "Top 5 in '95," "#1 in 4 years," "40% reduction in recordables each year for 5 years," and so on. What does all this mean? Usually very little.

Occasionally the mere emphasis on safety achieves it, and sometimes too soon. The "Top 5 in '95" goal was reached in 1992. How did the company do it? They do not know. Will the success last? No. At best there will be a "Hawthorne effect" or at worst hiding of accidents due to messages from above.

Assessing where we are now in our safety system is not easy; and, as we are learning every day, it is not done the way we thought it was.

In the early days of safety it was simple—we just looked at the statistics; a 0.5 accident frequency rate meant we were good and that our safety program was effective, whereas a 15.0 frequency rate meant that our program was not effective. It was simple; the frequency rate (of recordables or lost times) told us whether or not everything was working. And this was true until we started looking at the real truth—at the meaninglessness of the number we were using.

Do accident statistics reflect safety system effectiveness? Almost always they do not, unless you have many accidents to provide you with statistical validity. Without these numbers you are simply measuring how lucky or unlucky a unit has been. A supervisor of ten people can do absolutely nothing all year and attain a 0.0 frequency rate. Using a statistics measure to judge his or her performance is ludicrous.

At what point do accidents become a valid measure to judge performance? Actuaries state that only when about 1,129 accidents had occurred would they begin to judge the unit that generated those 1,129 to be so believable that future rates could be based on those figures. Rating bureaus recognize this by using a national figure when state figures are deemed to be too small for credibility. We all know this, yet we continue to judge supervisory performance by the accident rate on a month to month basis. Would it make any sense to judge supervisory safety performance (did the supervisor do anything to prevent accidents?) by a fatality rate each month (the supervisor still has the same ten people who were there last month and thus has done all of the necessary things)? Obviously this measure makes no sense whatsoever. Using a frequency rate is a touch less ridiculous than this (but not much).

So what do we use to measure the safety performance of line managers? Almost anything *but* accident statistics. How about an activity measure: did the manager do those defined activities that he or she agreed to? The activity measure would at least ensure some validity in the measurement.

System Approaches

Audits (the preidentification of what should be in place and checking to ensure that they are) are an integral part of safety management. We safety professionals have believed in audits since the 1950s. We have developed them into an art form. We have audits (constructed internally or externally), profiles, quantified audits, audits that end up in recommendations, and so on.

What good are they? Every large company does them. Most consultants sell them. Some safety programs are built around them.

Are they of value? No one knows. Do they predict safety system future results? Probably not. Do they forecast future serious incidents? Almost assuredly not. Then of what use are they? We do not know although, because most audits are regulation-oriented, they might fend off fines. Do they? Though we do not know, we assume that they do. Does regulatory compliance get fewer accidents? We are almost sure it does not. So why do we perform audits? For regulatory compliance. You should aim them that way, do them that way, and not confuse regulatory compliance (a must) with safety (a must).

All this is *not* to tell you not to do audits. It is simply to say that audits are good for some things, such as regulatory compliance, but not worthwhile for other things, such as a safety system effectiveness assessment (unless the audit format is tailored to that purpose).

Here we always run into serious objections from managers, executives, and particularly safety professionals, who cite these (and other) reasons for using frequency rates to measure safety effectiveness:

- Why, OSHA requires it.
- Why, it is our only bottom-line measure.

- Why, we have always done it that way.
- Why, Heinrich said so.

At this point you have to make crucial decisions: Do you want to control losses, or have a "safety program"? Do you want to comply with regulatory guidelines, or to stop people from getting hurt? Do you want to have a safety system that is in step or out of step with your management's and current management's thinking? All of these questions can be answered either way; it is your choice.

Generally, our approaches to the analysis of safety performance range from the simple to the complex. Included are such traditional approaches as accident investigations, inspection, job safety analysis, and so forth, and such nontraditional approaches as fault tree analysis, man safety analysis, and so on. Perhaps a logical starting point is to look at the types of analysis models that are available to us from management theory. William Johnson, in *MORT—Management Oversight Risk Tree*, discusses a variety of management methods relevant to the safety system (see Figure 2-1). He suggests that a repertory of tested management practices, rather than only one approach, is needed to meet the complex needs of safety.

Johnson believes that it is probably not as important which management concepts are used as that some be chosen explicitly and used. It is only when some reasonably firm approach is well known and used that it is possible to (1) hold personnel to any standard of analysis and (2) judge the value of any standard or practice.

Johnson discusses a selection of management methods particularly relevant to safety (see Figure 2-1). Juran's control cycle is the first approach of the second row in Figure 2-1. This, as with most of the models in the figure, is self-explanatory. The Kepner-Tregoe system, shown next, is particularly appropriate for a safety system in both language and in logic; for example, probable cause, minimum threat, problem probability and seriousness, preventive action, trigger-contingent action, and so on.

The Kepner-Tregoe sequence can also be seen as a simplified cycle, which can be superimposed on the basic safety system for comparison purposes. In general, the Juran and Kepner-Tregoe techniques seem to have more value for safety than other management approaches.

MORT suggests several approaches to error reduction, such as the Improving Human Performance model. The Professional Manager method, while useful, is perhaps less explicit for safety than the first three models of Figure 2-1.

Management By Objectives (MBO) is quite useful in safety and can augment one or more of the problem-solving processes. The traditional delegation of responsibility, authority, and resources is the main approach, and should remain so. However, there are weaknesses as well as strengths. These weaknesses can be offset by a full repertory of management methods.

Johnson, in *MORT*, also provides a model of the safety system (performance

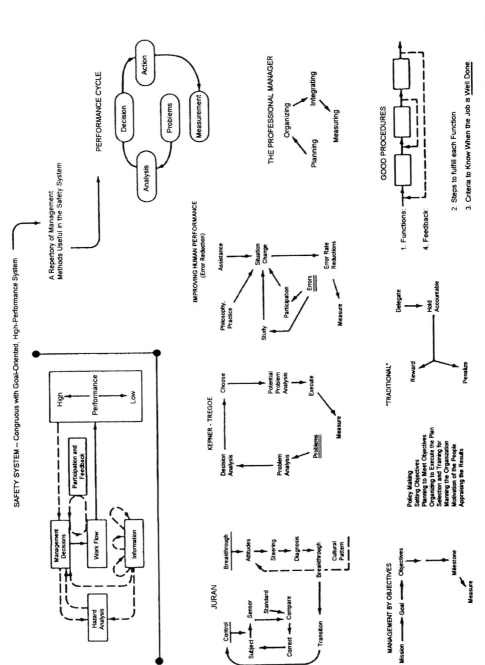

Figure 2-1. Management approaches. (From W. G. Johnson, *MORT—Management Oversight Risk Tree*, Washington, DC: U.S. Government Print-

How We Traditionally Have Analyzed   19

Figure 2-2. Nertney hazard analysis wheel. (From W. G. Johnson, *MORT*, Washington, DC: U.S. Government Printing Office, 1973.)

cycle) and indicates that it is based on the normal management performance cycle of problem–analysis–decision–action–measurement.

Based upon these kinds of management approaches, each safety professional must select his or her own approach. Johnson describes several, including the Nertney Hazard Analysis Wheel (Figure 2-2) and the General System for Assessing Organizational Risk (Figure 2-3). They are included as examples of approaches used by different organizations. Johnson also includes a model of the basic essentials of the U.S. Steel safety program (Figure 2-4). For further information on any of these, see *MORT*.

Figure 2-5 depicts an analytical system for loss control developed by L. H. Dawson, former corporate safety director of Kodak. In the figure, most of the aspects of the scope and functions are depicted. The figure shows the methods available for analysis (audits, process reviews, job safety analyses, accident investigations, and inspections) as well as the methods available for improvement

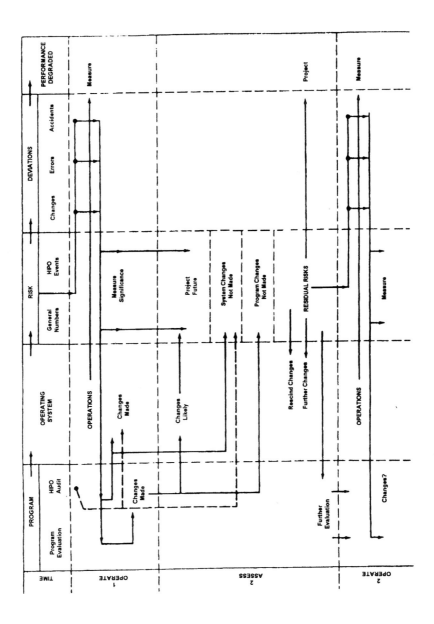

Figure 2-3. General system for assessing organizational risk. (From W. G. Johnson, *MORT*, Washington, DC: U.S. Government Printing Office,

How We Traditionally Have Analyzed 21

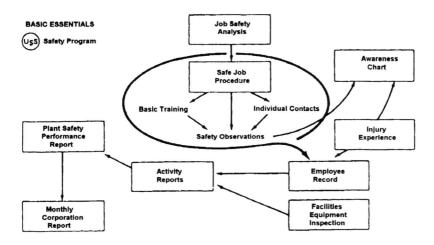

Figure 2-4. U.S. Steel's safety program. (From W. G. Johnson, *MORT,* Washington, DC: U.S. Government Printing Office, 1973.)

(inform/train/educate, design/modify/implement change, correct performance/ change human environment).

There thus seem to be enough different approaches available to us in our analysis. The first step in the analysis is the gathering of information, finding out what currently exists in the organization. Here also there are many possible methods available to us, from the simple to the complex.

The simplest approach is probably the checklist. We simply develop (or borrow from someone else) a long list of things that are important and relevant to what we are looking at, in this case those things important to help management in controlling accidental loss.

Besides the checklist approach to obtaining information, we have traditional approaches of inspection and the investigation of accidents. Both provide us with considerable information about certain aspects of our safety approaches. There also is the interview and questionnaire approach to obtaining information from employees and others in our organizations, about the safety program and its acceptance and value.

The various systems safety approaches are also available as systematic approaches to information gathering and analysis. One of the more comprehensive of these is *MORT*. Figure 2-6 shows the kind of analysis of a project or system completed by *MORT*.

MORT discusses in some detail the analysis of project or system risk and suggests these basic elements (see Figure 2-7):

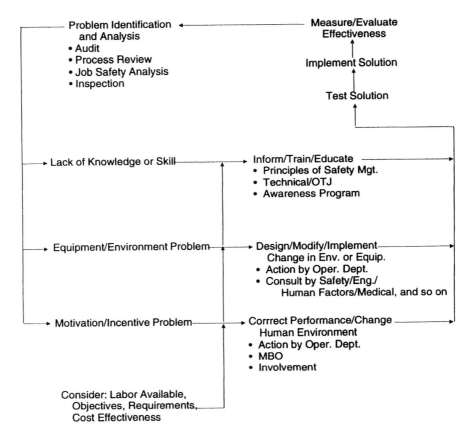

Figure 2-5. Analytical system for loss control. (From L. H. Dawson, Rochester, NY: Eastman Kodak Company, 1978.)

1. A choice of alternatives.
2. Identification of hazards, and the probability and consequences of accidents.
3. Comparison of the likely outcomes against criteria.

Criteria include not only the organization's safety goals but also cost, scheduling, reliability, quality, and other internal criteria, plus employee and public concerns and constraints.

The *MORT* charts use a risk assessment system with the following elements:

1. Comparison with goals
2. Experience data

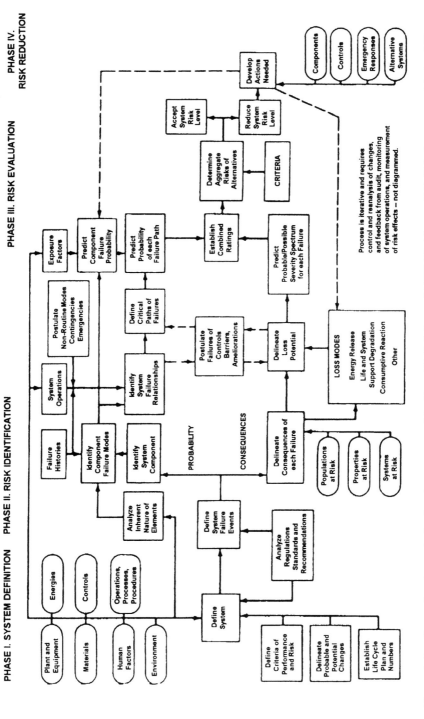

Figure 2-6. Comprehensive analysis of project or system risk. (From W. G. Johnson, *MORT*, Washington, DC: U.S. Government Printing Office, 1973.)

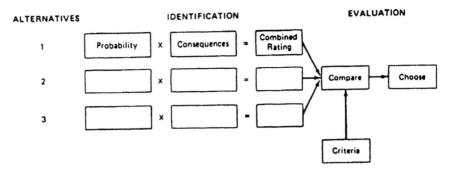

Figure 2-7. Basic elements of risk assessment as per *MORT*.

      (a) Rates.
      (b) Causes and circumstances.
  3. Hazard review process with three factors:
      (a) Triggers to activate hazard analysis.
      (b) Knowledge and information on hazard reduction.
      (c) An adequate hazard analysis process definition showing analytic operations to be performed.
  4. Safety program review

The *primary objectives* of a risk assessment system should be to provide a manager (at any level) with the information needed to (1) assess residual risk and (2) take appropriate action, if he or she finds residual risk unacceptable. Important but *secondary objectives* would include (3) comparative evaluation of two or more units and (4) research evaluation of two or more hazard reduction schemes to add to the state-of-the-art.

As management has the legal and moral responsibility for safety, it seems that safety information and measurement programs should be primarily designed to answer these critical safety questions of management:

1. What are the nature and magnitude of the organization's accident potentials?
2. What has been done to reduce risk?
3. What is the long-term level of residual risk?
4. What additional measures to reduce risk have been considered and rejected on "practical" (investment/benefit/value) grounds?
5. Are the safety programs actually operating as described in manuals and procedures?

AREAS TO ANALYZE

Once you have determined an approach to analysis, it is necessary to make a determination of what exactly must be analyzed, which areas are relevant to results. In our case the areas seem to fall into some specific and distinct categories, some relating to physical conditions and some to the management system that is there to control, and others relating to the behavioral or psychological climate that exists to influence the work force.

Some research has taken place in this area, which also might suggest the areas to analyze. R. Simonds and Y. Shafai-Sahrai reported on some research where they matched 11 pairs of firms selected so that the two members of each pair were each in the same state, in the same industry, and of approximately the same size but with markedly different injury frequency rates. These firms were studied for the same period of time. They reported the following:[1]

> All were visited and their records analyzed to see whether differences in management involvement in safety, promotional efforts toward safety, work force characteristics, or physical conditions were related with the better or poorer injury records. Factors found to be related to lower injury frequency rates were: (1) Top management involvement in safety; (2) better injury recordkeeping systems; (3) use of accident cost analysis; (4) smaller spans of control at foreman level; (5) recreational programs for employees; (6) higher average age of employees; (7) higher percentage of married workers; (8) longer average length of employment with the company; (9) roomy and clean shop environment; and (10) more and better safety devices on machinery. Factors not related [to] injury frequency were: (1) Efforts to promote safety through workers' families, and (2) quality and quantity of safety rules.

Studies like this should provide some insights into the areas of management relevant to safety success in an organization.

Dr. F. Rinefort completed a similar study offering insight into the factors related to accidents in an organization. His study consisted of a detailed review of the safety activities and the safety performance of 140 Texas chemical, paper, and wood product manufacturers, with a statistical analysis of the information obtained.

Dr. Rinefort prepared a questionnaire that asked detailed questions about top management safety activities, safety and health staff, new employee orientation and training, safety rules, activities to maintain interest in safety, safety meetings, safety inspections, the provision of personal protective equipment, the correction

---

1. From R. H. Simonds and Y. Safai-Sarai, Factors apparently affecting injury frequency in eleven matched pairs of companies, *Journal of Safety Research,* September 1977.

of unsafe physical conditions, physical examinations, injury treatment facilities and staff, off-the-job safety activities, safety training for experienced employees, and safety recordkeeping activities.

His questionnaire also asked respondents to indicate[2]

> the relative effectiveness of these various activities by providing a rank ordering of them. Another portion of the questionnaire asked about the interest of top management, the design and layout of the facility, and about housekeeping at the location.
>
> Measures of the safety record or performance were obtained by questions about the number of fatalities, lost-time injuries, days lost, medical cases, first-aid cases, man-hours worked, injury frequency rates, injury severity rates, and dollar costs of work injuries.

The monetary costs of the safety activities were calculated in terms of dollars per employee (see Figure 2-8). The estimated monetary cost of work injuries to Texas firms was also calculated, which consisted of the direct or insured cost of work injuries, an estimate of the monies paid beyond these direct costs to Texas worker's compensation insurance carriers, and an estimate of the indirect or noninsured costs of work injuries (see Figure 2-9).

Dr. Rinefort analyzed the data in three ways. First, graphic comparisons were made. Second, multiple linear regressions were computed (see Figure 2-10). Third, costs were calculated as percentages of the average annual hourly payroll.

Dr. Rinefort's analysis indicated that those activities that[3]

> together make up a safety program explained approximately 60% of the difference between the cost of work injuries in large or medium-sized firms or in firms with more than 50 employees. Approximately 27% more of this difference was explained by the size of the firm as measured by the number of employees and by span of control of supervisors per employee measured here in terms of the cost of such supervision per employee. Therefore, for firms of these sizes, appropriate safety programs can reduce work injury costs but only up to a point. In small firms or those firms with fewer than 50 employees, safety program activities explained 75% to 95% of the difference between the cost of work injuries for firms with low injury costs and those with high injury costs.

Dr. Rinefort's review of the data indicates[4]

> that the 14 safety activities, the span of control or available supervision, and the firm size as measured by the number of employees interact with each

---

2. From F. Rinefort, A new look at occupational safety, *Professional Safety*, September 1977.
3. Ibid.
4. Ibid.

How We Traditionally Have Analyzed   27

	Chemicals		Paper Products		Wood Products	
	Low Injury Freq. Rate	High Injury Freq. Rate	Low Injury Freq. Rate	High Injury Freq. Rate	Low Injury Freq. Rate	High Injury Freq. Rate
Small firms	$414	$645	$280	$346	$151	$285
Medium-sized firms	493	624	187	158	111	221
Large firms	757	720	203	525	146	222

Figure 2-8. 1974 cost of safety activities for Texas chemicals, paper-products, and wood products manufacturing firm respondents. (From F. Rinefort, A new look at occupational safety, *Professional Safety*, September 1977.)

other in complex, subtle ways to explain differences between those firms with low work injury costs and those with high injury costs. The study further shows that there are no easy, simple answers to the question of how to best reduce work injury costs but that relationships between the cost of safety activities and work injury costs do exist and that they can be estimated. It was found that a better combination or mix of these various safety activities rather than greater monetary expenditures for some of them was frequently the best way to reduce work injury costs. This better combination seemed to consist of sufficient expenditure for safety rules, off-the-job safety, safety training, safety orientation, safety meetings, medical facilities and staff, and of good top management interest and participation in safety. The most effective mix, therefore, would seem to be a balanced approach which combines both engineering and non-engineering and which probably places more emphasis upon the non-engineering aspects.

	Chemicals		Paper Products		Wood Products	
	Low Injury Freq. Rate	High Injury Freq. Rate	Low Injury Freq. Rate	High Injury Freq. Rate	Low Injury Freq. Rate	High Injury Freq. Rate
Small firms	$16	$475	$125	$435	$ 36	$1192
Medium-sized firms	51	335	144	631	388	1169
Large firms	33	318	100	333	177	844

Low injury frequency rate refers to firms which experienced work injury frequency rates lower than the average for the industry.
High injury frequency rate refers to firms which experienced work injury frequency rates higher than the average for the industry.
Small firms employed fewer than 50 full-time employees. Medium-sized firms employed between 50 and 199 full-time equivalent employees. Large firms employed 200 or more full-time equivalent employees.

Figure 2-9. 1974 cost of work injuries per employee for Texas chemicals, paper products, and wood products manufacturing firm respondents. (From F. Rinefort, A new look at professional safety, *Professional Safety*, September 1977.)

Size of Firm	Chemical	Paper	Wood Products
Small	Off-the-job	Off-the-job	Orientation
	Records	Physicals	Training
	Training	Orientation	Meetings
	Rules	Staff	Management
	Inspections	Span control	Span control
	Guarding	Guarding	Rules
	Management	Inspections	Equipment
	Staff	Meetings	Physicals
	Orientation	Management	Interest
	Equipment	Medical	
	Physicals	Records	
Medium	Management	Medical	Rules
	Off-the-job	Training	Training
	Rules	Equipment	Management
	Medical	Interest	Medical
	Equipment	Orientation	Staff
	Training	Span control	Interest
	Meetings	Guarding	Guarding
	Span control	Inspections	Meetings
	Orientation	Physicals	Orientation
	Interest	Management	Physicals
	Physicals	Rules	Records
	Records		
Large	Off-the-job	Off-the-job	Medical
	Physicals	Records	Records
	Inspections	Interest	Interest
	Medical	Inspections	Span control
	Training	Meetings	Meetings
	Span control	Training	Guarding
	Guarding	Span control	Physicals
	Staff	Staff	Orientation
	Orientation	Orientation	Staff
	Records	Management	Training
	Interest		Equipment
	Management		Rules
			Management

The activities listed above the horizontal line in each grouping were cost effective. The activities below the horizontal line were cost ineffective.

Figure 2-10. Decreasing rank order of the cost effectiveness of selected variables upon the cost of work injuries per employee for Texas chemical, paper, and wood product firms based upon a multiple linear regression analysis. (From F. Rinefort, A new look at professional safety, *Professional Safety, September 1977.*)

Studies in the nature of Rinefort's provide valuable information to the practicing safety professional about which areas to analyze and emphasize.

BENCHMARKING

It has been particularly popular in recent years for organizations to benchmark, to look at other organizations to see what they are doing in safety (etc.). Benchmarking is regarded as a good tool, perhaps in the belief that if we take such a look, it will provide us with some keys to safety excellence.

Does the process help? Probably only those who have done it can answer the question. I have worked with companies that have considered it extremely helpful. Others have found it a waste of time. An example of a benchmarking study follows.

Best in Class Benchmarking Survey: Summary of Results

Background

Eighteen companies were evaluated in four areas:

1. Safety leadership—accountability/ownership.
2. Safety performance—incentive/recognition/counseling.
3. Communication/sharing of information.
4. Training for safety.

Best Practices

Safety Leadership
   a. Roles and responsibilities should be clearly defined for everyone from the CEO to the line supervisor. The level of responsibility for which these individuals are held accountable must reflect their authority to effect change.
   b. For each important corporate safety initiative, identify a safety champion among top corporate managers and task that person with driving that initiative to complete implementation across the corporation.
   c. Ownership by senior management of safety programs and performance should be visible to the organization. Active accountability can be demonstrated through discussion of performance with peers at management review meetings and supporting personnel at "town meetings." Senior

management must be accountable for safety performance of their sites or business units through their operating budgets (including funds available for compensation).
d. The best site programs should be identified and consistently applied across the corporation. Identification through internal "islands of excellence" and application through a seed program are recommended.

Safety Performance—Incentive/Recognition/Counseling

a. Leaders in safety performance use metrics that effectively drive their continuous improvement efforts. Leading indicators (e.g., observations) are used to predict changes in safety performance. Monitor safety performance versus program implementation at all sites.
b. Performance targets are well defined and clearly communicated. The targets challenge the organization to continuously improve. Best companies review and reset expectations according to feedback from employees, managers, the performance metrics, and their competition. Safety performance expectations should be in harmony with business objectives.
c. Safety performance is rewarded and tied to compensation and/or operating budgets. Tying safety performance to bonuses and merit pay is consistent among all leaders in safety performance.
d. Discipline is assessed consistently according to well-developed guidelines. Leading companies include the results of accident investigation procedures such as root cause analysis in formulating disciplinary actions.

Communicating/Sharing of Information

a. Companies that are leaders in safety communication solicit feedback from the recipients of safety information; they put senior managers on the plant floor to ask questions and communicate one-on-one about safety. The leaders discuss their safety programs and company safety performance expectations with prospective employees.
b. Leading companies use a variety of media to deliver timely and consistent messages throughout the organization. Successes are communicated with the same commitment and enthusiasm as accidents and incidents.

Training for Safety

a. Training requirements are tracked to determine the status of training needs and level of compliance.
b. The leader in training solicits feedback from course attendees and the course sponsor. Safety training is evaluated at the time of information delivery as well as after the fact.
c. Behavior observations are used to indicate whether attendees retained the information presented in the training program. Observations are documented and used to modify the training program if needed.

How We Traditionally Have Analyzed 31

d. Innovative training techniques are used that maximize the potential for learning, such as computer-based, self-paced training and trainee involvement through re-creations of incidents.

The remainder of this book will look at the various areas that you may choose to include in your analysis approach.

Chapter *3*

Analysis by Workers

As indicated in the last chapter, analysis of safety system effectiveness by accident statistics is usually a waste of time. Because of the sample size, they are usually invalid, mostly measuring luck. Analysis by audit might be equally invalid, measuring regulatory compliance more than safety system effectiveness.

This chapter discusses those measures that do have (or can have) statistical validity: perception surveys, interviews, and behavior sampling.

THE PERCEPTION SURVEY

The perception survey is used to assess the current status of an organization's safety culture. Critical safety issues are identified, and any differences in management and employee views on the effectiveness of company safety programs are clearly demonstrated.

There are a number of safety perception surveys on the market today. The survey described here is one that I helped to develop and use in my consulting work, and therefore the one I know best. It is available commercially as a part of the video series "The Challenge of Change," produced by Core Media, Inc. of Portland, Oregon.

The survey begins with a short set of demographic questions that can be used to organize the graphs and tables that show the results. Typically responders are asked about their employee level, their general work location, and perhaps their trade group. At no point is an employee asked questions that would enable him or

33

her to be identified by the people scoring the results. Complete anonymity is critical to the success of the survey.

The second part of the survey consists of 100 questions, which are designed to uncover employee perceptions about 21 safety categories.

All questions are answered with either "yes" or "no." Employees also have the option of leaving the answer blank. Whether a "yes" answer or a "no" answer is considered positive or negative depends upon the individual question.

To score the survey, the number of positive answers for each question is determined and divided by the total number of responses to the question. This computation provides a percent positive response.

Each question may impact the score of more than one category. A cumulative percent positive response is computed for each category. The percentages for the categories are graphed (see Figure 3-1) to display the results in descending order of positive perception by the line workers. Those categories on the right-hand side of the graph are the ones that are perceived by employees as being the least positive and therefore are the most in need of improvement.

Examples of perception survey results are shown in Figures 3-2, 3-3, and 3-4.

The easiest way to implement the survey is to set up the software on a personal computer and have each person take the test on the computer. The computer can then process the data and print the appropriate graphs and reports.

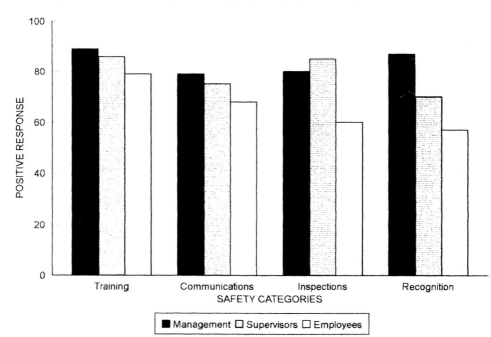

Figure 3-1. Example of perception survey results.

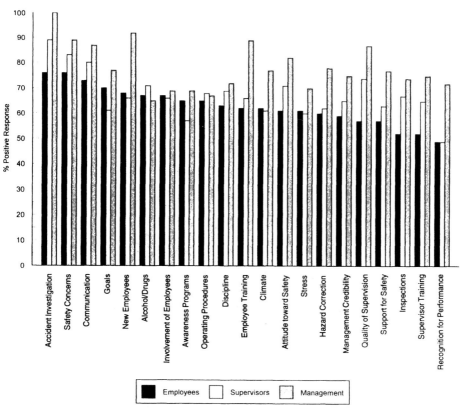

Figure 3-2. Responses from one location of a steel company.

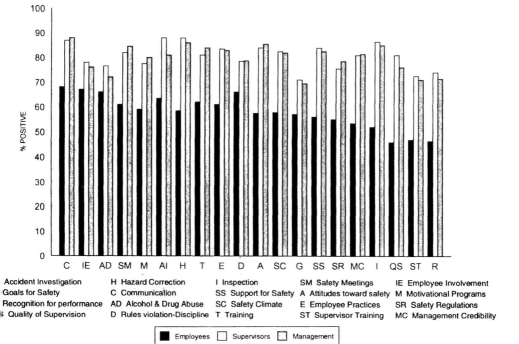

Figure 3-3. Responses from one department of a major railroad.

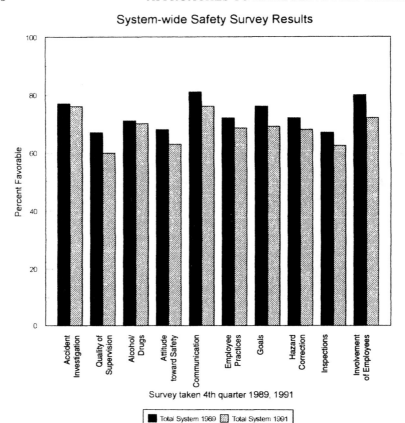

Figure 3-4. Comparison year to year of a major railroad (only ten categories shown).

The second alternative is to print all of the survey questions on paper forms and have employees fill them out by hand. When they are done, the form can be returned for the results to be entered into the computer program. These two alternatives also can be combined if, for example, local employees will use the computer, while paper forms are sent to employees at remote sites.

The categories measured by this perception survey are shown in Figure 3-5.

INTERVIEWING

The interview method is a good alternative to the perception survey. Some companies even use both methods—the survey for the big picture, the interview for specific details. (See below, discussion of which method to use.)

Analysis By Workers

The Perception Survey: 21 Safety Categories

This sample questionnaire may be useful to anonymously assess workers' perceptions of their workplace's safety culture. Once employers have real data, work groups composed from every echelon of the organization can work to improve the areas that may need it.

1. Accident Investigation — Does your safety system deal positively with the investigation of accidents?
2. Quality of Supervision — Are supervisors perceived to be competent in accident prevention?
3. Alcohol and Drug Abuse — Are employees with substance abuse problems allowed in the workplace?
4. Attitudes Toward Safety — Is there a positive attitude toward safety at all levels of the organization?
5. Communication — Do managers and employees communicate freely on safety issues?
6. New Employees — Are new employees thoroughly trained in safety?
7. Goals for Safety Performance — Do workers and management meet together to formulate behavior oriented safety goals?
8. Hazard Correction — Is there an effective system for dealing with reported hazards?
9. Inspections — Are there regular inspections of all operations?
10. Involvement of Employees — Are there opportunities for employees to become involved in safety through such means as quality improvement teams, ad hoc committees or effective supervision?
11. Awareness Programs — Do you hold awareness programs that stress safety both on and off the job?
12. Recognition for Performance — Is good safety performance recognized at all levels of the organization?
13. Discipline — Is the company perceived as taking a fair approach to handling rules and infractions?
14. Safety Contacts — Are there regular safety contacts with all employees?
15. Operating Procedures — Are safe procedures seen as both necessary and adequate by all levels of the organization?
16. Supervisor Training — Are supervisors perceived as well trained and able to handle problems related to safety?
17. Support for Safety — Is the whole organization seen as working together to create a safe work environment?
18. Employee Training — Do employees feel that they receive adequate training in how to work safely?
19. Safety Climate — Has a climate been created that is conducive to adopting safe attitudes and work habits?
20. Management Credibility — Is management seen as wanting safe performance?
21. Stress — Does the organization have an abundance of stress claims?

Figure 3-5. Safety categories.

Like the perception survey, the interview method provides a means to prioritize which of the safety categories require attention in the company.

In smaller companies where there are only a few people to interview, the interview task force can conduct all the interviews by itself. In larger companies it may be necessary to recruit additional people. In either case, the interviewers should be outgoing, enjoy talking with people, easily create a sense of trust, and really listen to what other people are saying.

Using good questions does not influence the response of the interviewees, but rather lets them explain in their own words what they believe to be the most serious safety issues at the company.

Specific questions might be:

- What are the top three things that should be done around here to improve safety?
- What would you do around here if you were in charge?
- What do you think are the most important safety issues at this company?

The number of people to interview usually becomes apparent during the course of the interviews. As long as you are obtaining new information that the previous interviewees did not tell you, continue to interview more people. When it becomes clear that you are hearing the same information over and over again, then you have talked with enough people.

Conducting Interviews

First, discuss the purpose of the interview. Take a few minutes to explain to the interviewee that all he or she must do is answer to the best of his or her ability. Explain that the responses will be completely anonymous. The interviewee should understand that this is not a test, and no special knowledge is required.

Second, collect any required demographic data that will be used to categorize the responses. As with the perception survey, you will want to know the employee type, the location, and perhaps the trade group.

Third, find out the interviewee's major safety concerns by using open-ended interview questions. Allow sufficient time for the interviewee fully to develop his or her answers.

Fourth, after the interviewees have listed their top safety concerns, then more specific questions can be used to generate follow-up information about their responses or about issues that might have been raised during other interviews. This is the time to get at the details and causes behind the general issues.

Fifth, provide a summary of what you have heard, and give the interviewee a chance to correct any misimpressions.

The last step is to finalize the chart and identify which safety issues should be addressed (Figure 3-6).

Analysis By Workers

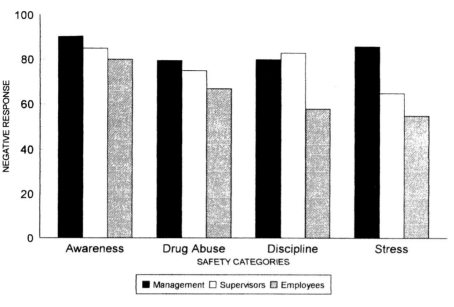

Figure 3-6. Interview method analysis chart.

Interviewing Skills

A number of skills are required for one successfully to accomplish the interview method.

Solid interviewing techniques are extremely important. Without the structure of the perception survey, each interviewer is essentially on his or her own during the interview process. Although the basic questions should be written down as part of an interview schedule, the interviewers must be free to use their own discretion on the follow-up questions.

The interview process is labor-intensive. Careful time management is critical for the interviews to be completed in a reasonable amount of time. You need to make sure that interviews are conducted efficiently and that you do not interview more people than you need to achieve your goals.

Good listening skills are the most important element in the entire process. People frequently tend to hear what they expect to hear rather than what the person is actually saying. All of your interviewers must have the ability truly to listen to people.

A feeling of trust also is critical for success. You are asking the interviewees to tell you about things that are wrong with your system. They will naturally fear that they might be punished in some way for talking about what their supervisors or other employees are not doing right. They must trust that the interviewer will keep their answers anonymous.

Perception Survey vs. Interviewing

Either the perception survey or the interview method can be used to assess your safety culture. Both will produce good results.

In most cases the perception survey should be considered first. It is a well-structured, statistically valid instrument based on many years of experience in the field. No special skills are required to conduct it, and a personal-computer–based software program is available to support the process. Several independent consultants are available to provide assistance, should it be desired. Because the survey uses a standardized form, it is quite easy to compare results at different sites or across different years.

The interview method should be considered if your company does not have easy access to computer support, or if for some reason surveys are not appropriate for your environment. It does not require the same amount of administration as the perception survey and usually does not require as many employees to participate. Further, it allows line workers to explain more about their views than could be obtained from a standard survey form. However, the interview method requires considerable skills on the part of the interviewers, and results may vary drastically, depending upon how well the interviewers are trained.

One excellent approach is to use the perception survey initially to assess the safety culture and then to train the problem-solving teams in the interview method so that they can investigate the key safety issues in greater depth.

SAFETY BEHAVIOR SAMPLING

Safety sampling measures the effectiveness of the line manager's safety activities, but not in terms of accidents. It measures his or her effectiveness before the fact of the accident by taking a periodic reading of how safely the employees are working.

Like all good accountability systems or measurement tools, safety sampling is an excellent motivational tool, for each line supervisor finds it important for his or her employees to be working as safely as possible when the sample is taken. To accomplish this, the supervisor must carry out some safety activities, such as training, supervising, inspecting, and disciplining.

In those organizations that have utilized safety sampling, many report a rather

Analysis By Workers

good improvement in their safety record as a result of the improved interest in safety on the part of line supervisors.

Safety sampling is based on the quality control principle of random sampling inspection, which is widely used by inspection departments to determine quality of production output without making 100 percent inspections. For many years industry has used this inspection technique, in which a random sampling of a number of objects is carefully inspected to determine the probable quality of the entire production. The degree of accuracy desired dictates the number of random items selected, which must be carefully inspected. The greater the number inspected, the greater the accuracy.

For details on how to conduct safety sampling, see Appendix B.

PART *II* | *The Areas to Analyze*

- In Part II we suggest a number of areas you may wish to analyze in your organization, areas that we believe to be germane to controlling loss. The 21 areas described may or may not be relevant to you. Our research has found that the 21 have been quite relevant to most situations. World class companies tend to be effective in these areas, but you must decide if they are the proper ones in your company.

    The 21 categories are in no way meant to suggest a "package program." The point here simply is that you should look at those 21 areas, and if you think (or your employees think) that there are weaknesses here, then fix those weaknesses in any way that seems appropriate for your environment. We offer no package solutions, but just some ways to help you to diagnose what you need to do—where your employees should concentrate their efforts—and come up with solutions to what needs fixing.

All 21 provide a look at your management system—what you are doing in each area that will control losses. These are the 21 areas, as grouped in the chapters that follow:

—*The management system you have in place to achieve continuous improvement of your safety process (Chapter 4), including:*
- Your accident investigation process.
- Whether your employees are involved.
- Your operating practices.
- Your discipline policies.

—*The management system you have in place to build a positive safety culture (Chapter 5), including:*
- What climate actually exists.
- Management's real credibility.
- Support for safety.
- The recognition the worker receives (or does not receive) for contributions.
- The general attitudes of your employees.
- The amount of stress that people feel daily on the job.

—*The management system you have in place to improve the skills of your supervisors and managers (Chapter 6), as evidenced by:*
- Effectiveness of your supervisory training.
- The perception of the quality of your supervisors.
- Your goal-setting process.

—*The management system you have in place to improve the skills of your employees (Chapter 7), including:*
- Their training.
- Your handling of the new worker.
- Your communication effectiveness.

—*The management system you have in place to improve employee behavior (Chapter 8), including:*
- Your safety contacts.
- Your alcohol and drug program(s).
- Your awareness program(s).

The Areas to Analyze

—*The management system you have in place to improve physical conditions (Chapter 9), including:*
- *Your inspections.*
- *Your hazard correction procedures.*

In this part of the book, we attempt to discuss each of the 21 areas. Again, all 21 may not be applicable to you, and there may be others more relevant. (See Chapter 10.) Our purpose here is to present a starting point, a way to assess your organization.

In each of the 21 areas we give some background covering today's safety management beliefs. We also provide one person's (the author's) perception of what is necessary to achieve excellence in each area. We do this through some simple fault tree analyses showing the author's perception of what is necessary. Change these diagrams to suit your company.

Also, for some areas we suggest that you use other problem-solving tools—fishbone diagrams, flow charts, and so on—in the hope that you will involve your employees in the process of improving any area that looks as if it needs help.

Chapter *4*

The Management System for Continuous Improvement

As we have suggested earlier, it is crucial to use upstream measures of the system rather than to judge progress with an annual look at a frequency rate, or some other (probably equally invalid) measure of system effectiveness. It is necessary to work on continuous improvement of the process of incident control.

Such analysis requires you to assess:

- Your accident investigation procedures.
- Whether your employees are involved.
- Your operating practices.
- Your discipline policies.

We look at each of these areas in this chapter.

YOUR ACCIDENT INVESTIGATION PROCEDURES

A good safety system deals positively with the investigation of incidents and thus minimizes any attempt to cover up causes and effects. In a good investigation process, the search for cause goes beyond identifying the unsafe act and condition; it searches for multiple causes, including weaknesses of the management system. It also searches for the reasons why unsafe acts are committed and the causes of human error. In addition, a good investigation process is not designed to find fault or assess blame. The purpose is to assist in continuous process improvement, to highlight system failures, and to discover what in the management system or climate is encouraging human error.

It used to be that the accepted model of accident causation was the old domino theory that said that accidents are caused by unsafe acts and/or unsafe conditions. We no longer believe that model; it has been replaced by new models based on research and 60 years of thinking beyond that of the old domino theory. These new models state that an injury or other type of financial loss is the result of an accident or incident, which is the result of (1) a systems failure and (2) human error.

An analysis of a systems failure answers these questions (and others):

- What is management's statement of policy on safety?
- Who is designated as responsible and to what degree?
- Who has what authority, and the authority to do what?
- Who is held accountable? How?
- How are those responsible measured for performance?
- What systems are used for inspections to find out what went wrong?
- What systems are used to correct things found wrong?
- How are new people oriented?
- Is sufficient training given?
- How are people selected?
- What are the standard operating procedures? What standards are used?
- How are hazards recognized?
- What records are kept, and how are they used?
- What is the medical program?

The second, and always present, aspect and cause of an incident or accident is human error. Human error results from one or a combination of three things: (1) overload, which is defined as a mismatch between a person's capacity and the load placed on that person in a state; (2) a decision to err; and (3) traps that are left for the worker in the workplace. We also must consider the culture of the organization as crucial to safety.

Overload

The human being cannot help but err when given a heavier workload than he or she has the capacity to handle. This overload can be physical, physiological, or psychological. To deal with overload as an accident cause, we have to look at an individual's capacity, workload, and current state. To deal with overload as an organizational cause, we have to identify the safety controls available for dealing with capacity, workload, and state.

A human being's capacity refers to physical and psychological endowments (what the person is naturally capable of); current physical condition (and physiological and psychological condition); current state of mind; current level of knowledge and skill relevant to the task at hand; and temporarily reduced capacity owing to drugs or alcohol use, pressure, fatigue, and so on.

Load refers to the task and what it takes physically, physiologically, and psychologically to perform it. Load also refers to the amount of information processing the person must do; the working environment; the amount of worry, stress, and other psychological pressure; and the person's home life and total life situation. Load refers to a person's level of motivation, attitude, and arousal and to his or her biorhythmic state.

Decision to Err

In some situations it seems logical to the worker to choose an unsafe act. Reasons for this might include:

1. The worker's current motivational field, in which it makes more sense to operate unsafely than safely. Peer pressure, the boss's pressure to produce, and many other factors might make unsafe behavior seem preferable.
2. The worker's mental condition, in which it serves him or her to have an accident.
3. The fact that the worker just does not believe he or she will have an accident (low perceived probability).

Traps

The traps that are left for the worker primarily involve human factor concepts. One such trap is incompatibility: the worker errs because the situation he or she works in is incompatible with the worker's physique or with what he or she is used to. A second trap is the design of the workplace (ergonomic factors).

Where Culture Fits

The most important consideration of all seems to be that everything we do to reduce human error and thus losses is totally dependent upon the employee's perception of the culture of the organization. The new models incorporate this as an overriding premise: culture is the key. Or, if you prefer, the "perception" of culture (which by definition is culture) is what makes or breaks safety; it is what makes any element of the process work or fail.

Management, through its vision, real values, systems of measurement and reward, and daily decisions, creates the culture of the organization. In that culture a number of processes and procedures attempt to operate. Their intent is to control losses. Some are overall (systemwide), such as whether or not people are held accountable for performance; and some are specific, such as the training a supervisor gets. These processes or procedures attempt to build a system that prevents loss, and to reduce human error caused by overload, deciding to err, and traps. When these procedures are successful, there are no incidents. When they fail, one occurs. The amount of loss is determined by the seriousness of the incident (largely luck) and by the systems we have in place to control the dollar outlay (light-duty programs, etc.).

To have any real effect, the organization's incident investigation procedure must look beyond unsafe acts and conditions into all of the above interrelated causes.

A model like those discussed above is shown in Figure 4-1.

Analysis Tools

There are certain essential tools to ensure investigation effectiveness, including fault trees, performance models, and fishbone diagrams.

Figure 4-2 is a simple fault tree, showing what must be present for an effective accident investigation process (author's opinion). For the reader not familiar with fault trees, a short explanation will suffice. There are two symbols on the tree: the AND symbol states that all of the things it connects must be present to achieve the desired result; the OR symbol states that only one of the conditions it connects is necessary to achieve success.

In Figure 4-2, the tree says that in order to have an effective accident investigation procedures, you must have;

      —An accurate assessment of the culture.
  AND—A real look for real causes, not symptoms.
  AND—Investigators properly trained in how to investigate.
  AND—A special investigation in process when indicated.
  AND—Follow-up on causes.
  AND—Communication of results to the proper people.

In the figure, the tree also states that to assess culture properly (the first box), you should have either a current perception survey or the results of current interviews. To do special investigations properly, you either must have specially trained people on the investigation process, or you must use special techniques to ensure in-depth investigations, such as TOR (Techniques of Operations Review) or CIT (Critical Incident Technique), either of which will force the investigator to look at underlying causes beyond the mere symptoms of unsafe acts and conditions.

If an individual is not doing the type of accident investigation desired, the performance model can be used to assess his or her performance (see Figure 4-3).

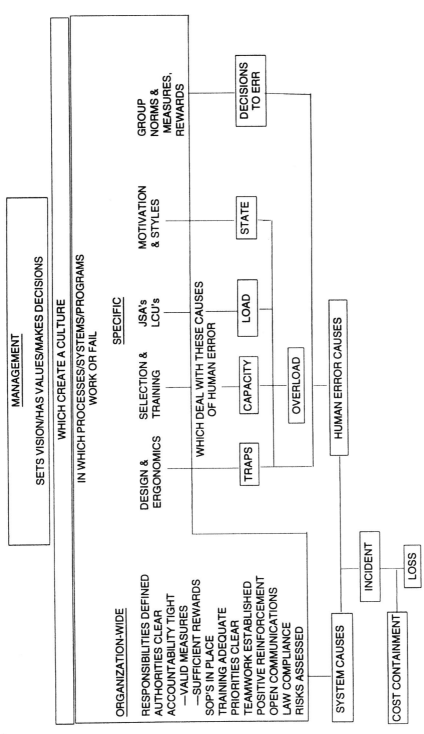

Figure 4-1. Accident causation model.

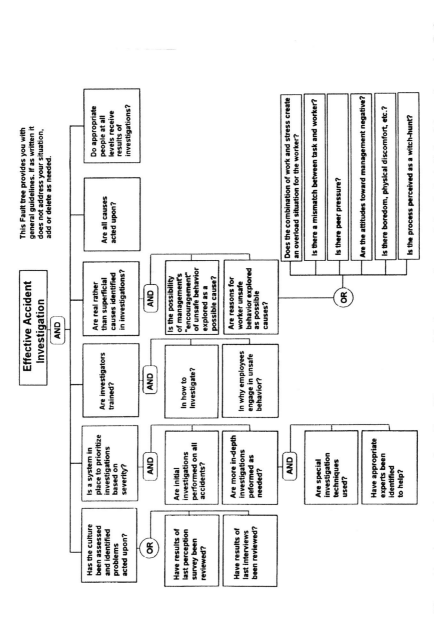

Figure 4-2. The accident investigation procedure. *Note:* The suggested fault tree can be amended to fit the beliefs of the reader, and to fit the needs of the organization.

The Management System for Continuous Improvement

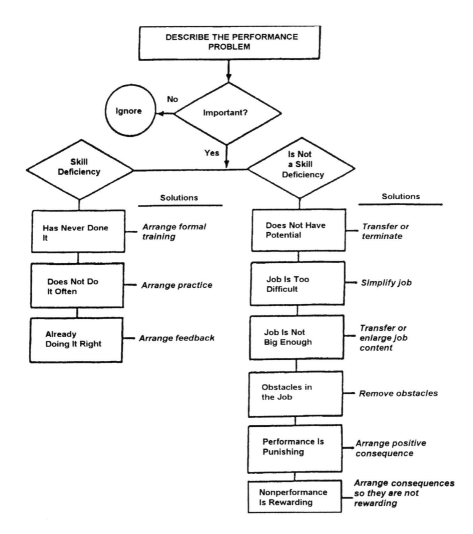

Figure 4-3. Performance model, for assessing an individual's performance.

Figure 4-4 shows a skeleton fishbone diagram. The fishbone can be used by your employees in brainstorming sessions to tap their knowledge and expertise in understanding why the process of accident investigation at your organization does not seem to be working well.

The fishbone diagram is a TQM (Total Quality Management) or an SPC (Statistical Process Control) tool commonly used by many companies to problem-solve with employee groups.

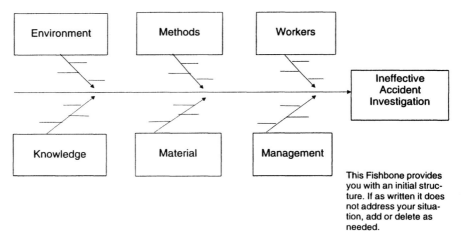

Figure 4-4. Fishbone diagram for ineffective accident investigation.

The fishbone diagram can be used as follows:

1. Select some element headings to get started (or principal causes).
2. List the principal causes on the horizontal lines and the subcauses on the diagonal lines. The horizontal center line or "bone" shows how the main bones or elements relate to the problem. At the right, you write the problem. It is in constant focus.
3. You can also use a fishbone diagram to depict how to achieve a goal. Substitute "goal" for "problem," and you can list all the factors for an effective improvement project.
4. You or the group can continually analyze the causes, including the apparent cause. Keep the diagram on display, as it can help in generating additional causes.
5. In addition, if you wish, you can rate each cause listed, from least likely to most likely cause, using a rating scale:

Rating Scale	Least Likely Cause			Maybe			Possible		Most Likely Cause	
	1	2	3	4	5	6	7	8	9	10

Then put the appropriate number (1 to 10) in front of each listed cause.

At this point you have identified the best place to start in the improvement process.

EMPLOYEE INVOLVEMENT AND PARTICIPATION

Good systems produce employee involvement in safety matters. Some observable results from getting employees involved are:

1. Personal interest in safety
   (a) Reminding others to work safely.
   (b) Correcting or reporting hazards.
   (c) Familiarity with safety record and goal setting.
2. Union support for safety programs (where applicable)
3. Active safety involvement teams
4. Ability of supervisor to reward good safety performance

There are many means of ensuring employee involvement in safety: quality improvement teams, quality circles, ad hoc committees, regular effective supervision, and many more. The real question is whether or not management has done things to ensure that any employee who chooses to can be involved in some meaningful activities to assist the safety effort. How many employees choose to be involved is a good reflection of the effectiveness of management's attempts.

The appeal of the participative approach was suggested as early as 1937 by H. Carey, who defined "consulting supervision" as the procedure whereby supervisors and executives consult with employees or their peers on matters affecting employees' welfare or interest prior to establishing policies or initiating action. From such beginnings, the concept of participative management has grown into a full-fledged approach to administration, affecting both patterns of organizational relationships and leadership style. The appeal of this concept is many-sided. Participation viewed in Carey's sense can be construed as a comfortable rationale for the paternalistic manager. Applied in the modern interpretation, it complements the political and social philosophies of democracy and individual self-actualization. Today psychologists believe that participation enhances the learning process and that a "democratic environment" may be particularly conducive to productive effort under certain conditions. These and other advantages enhance the value of the participative concept.

A few of the best studies of participation can be cited to show how the application of this technique can be used to further the adjustment of workers to changing conditions. The best-known studies are those carried on in the Harwood Manufacturing Company. In this company an attempt was made to come to grips with the workers' reactions to retraining. Harwood had severe turnover problems among workers forced to transfer from one job to another because of changing manufacturing needs. Many of these workers actually quit shortly after their retraining had begun. There were also a large number of voluntary layoffs during the period just before the workers reached factory standards of competence at their new jobs. In addition, retraining was slow and costly. Transferees took much longer to learn new jobs than did new workers, even those who had been highly skilled at their previous jobs.

A controlled experiment was set up in which one group of workers about to transfer was given routine information about the nature of the change and nothing else. Three other groups were asked to plan the way in which the factory would adapt to shifts in the market and in technology. These latter groups discovered through group discussion that their retraining was a necessity. The group that had merely been ordered to retrain showed high turnover and a long training period. The groups that had discussed the shift and arrived at their own decision about its necessity showed virtually no turnover and a phenomenally shortened training period. At a subsequent period, the original "no participation group" was reassembled and again retrained. This time they participated fully in the decision-making process. This study shows the significance of employee involvement.

Participation has been found to be successful in many situations. It does not always increase productivity (sometimes productivity increases, sometimes it stays the same, but it seldom declines), but it does improve attitudes, turnover, and morale. Workers do want to be involved in decisions where they believe they have a "right" to be involved. Workers' attitudes toward both company and union are strongly affected by the degree to which foremen and shop stewards welcome the participation of the rank-and-file workers in arriving at decisions. Involvement of all levels of employees in decision-making is a major factor in morale. In fact, the idea that it is a good thing to get employee participation in making changes has become almost axiomatic in management.

Participation, however, cannot be created artificially. We cannot buy it. We cannot hire industrial engineers, accountants, and other staff people who have the ability "to get participation" built into them. We cannot tell our supervisors and staff persons just to start participation. Participation is a feeling the people must have, not just an act of being called in to take part in discussions. People respond to the way they usually are treated, rather than to a strategy of sudden participation. Many supervisors have had unhappy experiences with executives who have read about participation and picked it up as a new psychological gimmick for getting other people to do what they want. Participation will never work as long as it is treated as a device to manipulate people. Participation must be based on respect, and respect is acquired when management faces the reality that it needs the contributions of the operating people.

Participative safety programs have been tried, and they work. One Midwestern company had these results:

> As this program developed employees recognized that they had a stake in safety, and we (the management) stayed away from telling them how to do

The Management System for Continuous Improvement 57

it. Then, instead of giving lip service to our enforced rules, employees reacted positively because the program for safety was theirs.

Another member of management stated:

Physically, little has changed; the machines, the protective devices, the products, the procedures, and most of the employees are the same as they were 18 months ago. What has changed is the attitude of the employees. Today, they are concerned with the safe method of performing the job, because they gain individually and collectively by being safety conscious. They also are aware that only their efforts can prevent accidents.

And the workers agree:

We had a safety committee for six years, but nobody took it to heart. Now we're thinking safety and quietly anticipating hazards before they create accidents.

In the safety program of the past, the closest thing we had to real participation was the safety committee. The pros and cons of safety committees have been aired for years without changing anyone's mind about them. They exist, and in some companies they fulfill a role of communication, quasi-participation, or, perhaps in some cases, real participation, for the few workers on the committee. The safety committee, however, is not what we are talking about for employees' participation in the safety decisions that pertain to their jobs.

The essentials to attaining real employee involvement are shown in the fault tree in Figure 4-5.

For brainstorming the causes of lack of involvement, you might use a fishbone diagram (Figure 4-6).

OPERATING PROCEDURES AND PRACTICES

Some companies call these procedures rules; others call them SOPs, SPIs (standard procedure instructions), or JSAs. This category refers to standards of work performance. Procedures can be set by staff, by top management, or by the employees themselves. However they are set, the procedures can be either meaningful tools or pages in a never-looked-at manual. This category looks at the perception of worthiness of these procedures.

Operating procedures are handled differently in different organizations. Usually they have been generated internally over time to direct workers and supervisors

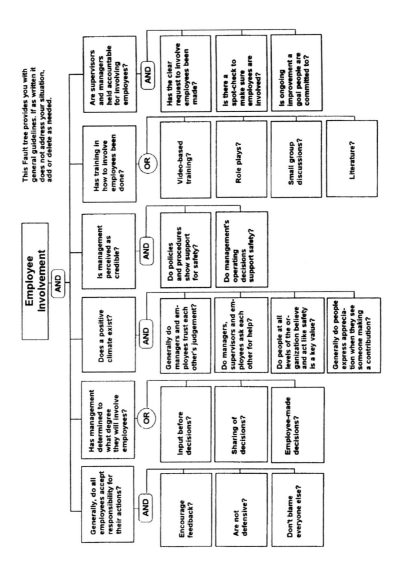

Figure 4-5. Employee involvement fault tree.

The Management System for Continuous Improvement 59

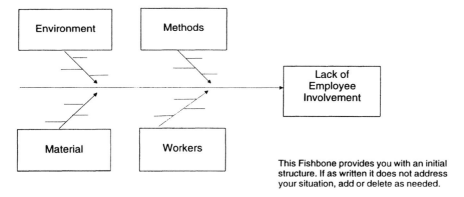

Figure 4-6. Fishbone diagram for lack of employee involvement.

into the safest way to do a job. Federal and state regulations have added to these procedures. When OSHA came into existence, for instance, they immediately required each company to incorporate hundreds of procedures in all kinds of operations above and beyond their physical condition standards. For example, they required, in 83 places in the regulations, that specific personnel be "instructed, trained, authorized, competent, licensed, responsible, or designated." They required, in 209 places, that specific inspections, tests, or records be kept. In 133 places they required that signs or warnings be required. Personal protective equipment is required throughout. And, these procedures have been added to extensively over the years.

It is important to ensure that the organization includes all of the regulatory requirements in its operating procedures. However, it is more important to recognize that having operating procedures does not mean that they will be followed, and to understand that even if they were followed, there would be accidents. Law compliance or procedure compliance does not ensure safety.

The essentials of effective procedures and practices are shown in the fault tree in Figure 4-7.

For brainstorming the causes of ineffective operating procedures, you might use the fishbone shown in Figure 4-8.

DISCIPLINE

Up to this point, we have not even mentioned enforcement and discipline, which is one of the cornerstones of traditional safety. Where does it fit in? Or does it? Perhaps the answer is "only as a last resort."

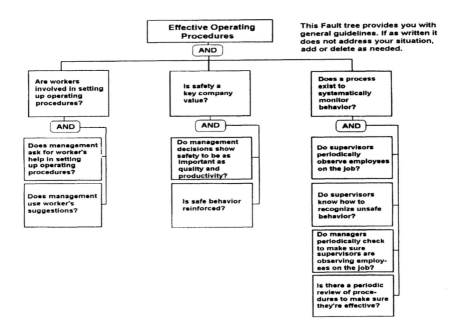

Figure 4-7. Effective operating procedures fault tree.

Although discipline rarely changes or improves behavior, a poor discipline system, or a good one poorly administered by a supervisor or a manager, can affect behavior adversely over the long haul. Indeed, research shows that positive reinforcement is much more powerful than negative reinforcement in both behavior-building and behavior-maintaining. This does not mean that discipline should be eliminated from our safety repertoire. It remains an important part, but it should be seldom used and only as a last resort. The day of the 600-page rule book and the issuing of tickets for violations probably is over. It is, in fact, counterproductive because it creates a "them versus us," a "police versus policed" environment. This is exactly the opposite of the team-building approaches of today.

Management, through its definition of policy, makes the decision that it wants safe performance from employees. However, management does not seem to be able to force safe performance even with the most sophisticated procedures now available. How can it make more effective use of discipline?

The Management System for Continuous Improvement

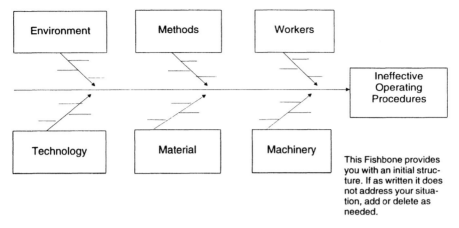

Figure 4-8. Fishbone diagram for ineffective operating procedures.

Unfortunately, the word discipline too often strikes a negative note in the minds of supervisors as well as employees. To the latter, it means rules that must be obeyed and penalties that are levied. To the former, it means a generally unpleasant task, the results of which almost always lead to antagonism. Yet discipline need not always be negative; it can and should be a positive process from both parties' standpoints.

Guidelines in Using Discipline

A good supervisor tries to create a climate in which his or her subordinates willingly abide by company rules. But even the best supervisor cannot expect perfection; rules still will be broken. What the supervisor does about these violations not only will influence the future behavior of the employee involved, but it can also have serious effects on the morale in his or her department. It can even affect future contract negotiations with a union. Here are some guidelines to help you use discipline effectively:

- *Know the rules, and make sure your subordinates know them.* You cannot maintain discipline unless you know what is allowable and what is not.
- *Do not ignore violations.* A supervisor does not have to issue a formal reprimand or disciplinary suspension every time a rule is broken. What he or she does will depend upon the nature and the circumstances of the violation and the employee's past record. The important point is that the supervisor must do something.

- *Get all the facts.* Most arbitrated disputes are over the facts of a discipline case. As soon as a supervisor believes there has been a violation, he or she should establish exactly what happened.
- *Choose the most appropriate disciplinary action.* Perhaps nothing puts a supervisor's judgment to the test more sharply than determining what discipline to give an employee who has violated a rule. The supervisor must draw a fine line between punishment that is too severe to be just and punishment that is too mild to be corrective.
- *Administer the discipline properly.* Telling an employee that he or she is being penalized for breaking a rule is no more pleasant for the supervisor than for the employee. This is the critical time for the supervisor to remember that the purpose of the discipline is corrective, not punitive.

The author of these guidelines also developed the following checklist for the supervisor to use when faced with a situation in which he or she must take disciplinary action:

1. Do I have the necessary facts?
    (a) Did the employee have an opportunity to tell his or her side of the story fully?
    (b) Did I check with the employee's immediate supervisor?
    (c) Did I investigate all other sources of information?
    (d) Did I hold my interviews privately to avoid embarrassing the employee?
    (e) Did I exert every possible effort to verify the information?
    (f) Have I shown any discrimination toward an individual or group?
    (g) Have I let personalities affect my decision?
2. Have I administered the corrective measure in the proper manner?
    (a) Did I consider whether it should be done individually or collectively?
    (b) Am I prepared to explain to the employee why the action is necessary?
        For instance:

Because of the effect of the violation on the employer, fellow employees, and himself or herself.

To help the employee improve his or her personal efficiency and that of the department.

    (c) Am I prepared to tell the employee how he or she can prevent a similar offense in the future?
    (d) Am I prepared to deal with any resentment the employee might show?

The Management System for Continuous Improvement 63

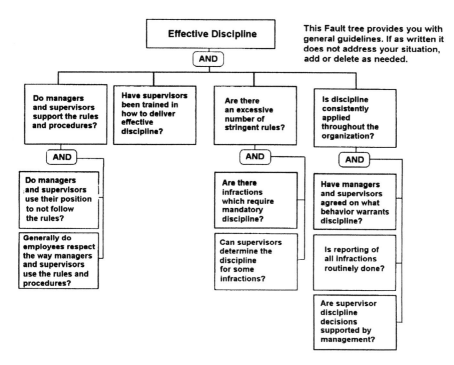

Figure 4-9. Effective discipline fault tree.

    (e) Have I filled out a memo for the employee's personnel folder or a letter describing the incident, to be signed by the employee? A copy of this memo or letter should be given to the employee, who should be told that he or she may respond in writing, for the record.

    (f) In determining the specific penalty, have I considered the seriousness of the employee's conduct in relation to his or her particular job and employment record?

    (g) Have I decided on the disciplinary action as a corrective measure, not a reprisal for an offense?

3. Have I done the necessary follow-up?
    (a) Has the measure had the desired effect on the employee?
    (b) Have I done everything possible to overcome any resentment?
    (c) Have I complimented him or her for good work?

(d) Has the action had the desired effect on other employees in the department?

A suggested fault tree for discipline is shown in Figure 4-9.

For brainstorming the causes of an effective discipline procedure, you might want to use a fishbone diagram.

Chapter 5

The Management System to Build Culture

There is nothing more important in the organization's safety system than the things the organization does to build or maintain a positive safety culture. This chapter discusses some of the "culture builders."

Actually all 21 safety categories are "culture builders"; if you do a good job on the 21, you will have the culture you want—you will have demonstrated that safety is a key value in the organization. The areas discussed here are particularly relevant.

SAFETY CLIMATE (OR CULTURE)

Although climate is difficult to define, it is easy to see and feel. The safety climate reflects whether or not safety is perceived by all to be a "key value" in the organization. The terms "climate" and "culture" are both used here. My favorite definition of culture ("culture is the way it is around here") reflects the unwritten rules of the ballgame that the organization is playing. Culture is what everybody knows; therefore, it does not have to be stated or written down.

The concept became a very popular management subject in the early 1980s, probably because of the popularity of *In Search of Excellence*, by Peters and Waterman. That book described what it was that accounted for the economic success of a number of companies. Other books followed, delving into the topic.

Basically, the concept of culture existed long before that book was published. Years earlier, Dr. Rensis Likert wrote *The Human Organization*, in which he described his research on "trying to" understand the difference in "styles" of different companies, and how these "styles" affected the bottom line. Dr. Likert coined the term "organizational climate." We now call it culture.

Likert not only researched climate; he also defined it in terms of ten characteristics:

1. Confidence and trust.
2. Interest in the subordinate's future.
3. Understanding of and the desire to help overcome problems.
4. Training and helping the subordinate to perform better.
5. Teaching subordinates how to solve problems rather than giving the answer.
6. Giving support by making available the required physical resources.
7. Communicating information that the subordinate must know to do the job, as well as information needed to identify more with the operation.
8. Seeking out and attempting to use ideas and opinions.
9. Approachability.
10. Crediting and recognizing accomplishments.

Likert invented a way to measure climate with a forced-choice questionnaire, which he administered to employees of an organization to determine their perception of how good the company was in the ten areas. He later took the perception survey results and ran correlational studies with such things as profitability, return on investment, growth, and other bottom-line figures, invariably getting extremely high positive correlations. Apparently climate determines results.

As culture (climate) became a popular management subject in the 1980s, executives began to look at their organizations and consider ways to "improve their culture." In many organizations one would find new posters on the walls described "their culture." We know today that if the management of a company must write it down and make a poster of it, they are not describing their culture—they are describing what they would like it to be. Nobody needs for the culture to be described—everybody knows what it "is like around here."

Culture and Safety

The field of safety generally ignored the concept of culture throughout the 1980s. As management attempted to improve culture through changing styles of leadership and through employee participation, safety professionals tended to change their approaches very little. They were using the same elements in their safety "programs" that they had always used. Safety programs typically consisted of the usual things: meetings, inspections, accident investigations, and using JSAs. These tools were perceived as the essential elements of a safety program. OSHA published a guideline in the 1980s suggesting that all companies should do all of these things. A number of states enacted laws requiring companies to do these same things. These elements were perceived as being a "safety program."

While OSHA and the states were going down the "essential element" track to safety (as was much of the safety profession), a number of research studies began to yield totally different answers to the safety problem. Most of the research results were consistent in saying that "there are no essential elements"; what works in one organization will not work in another. Each organization must determine for itself what will work for it. There are no magic pills. The answer seems to be clear; it is the culture of the organization that determines what will work in that organization.

Certain cultures do in fact include safety as one of their central values. Other cultures make it very clear that safety is unimportant. In the latter almost nothing will work; meetings will be boring, JSAs will be perceived only as paperwork, and so on.

What Culture Does

The culture of the organization sets the tone for everything in safety. In a positive safety culture, it says that everything done about safety is important. In a participative culture, the organization is saying to the worker, "We want and need your help." Some cultures urge creativity and innovation; some destroy it. Some cultures tap the employees for ideas and help; some force the employees never to use their brains at work.

What Sets the Culture

Let us mention just a few issues:

- How decisions are made: Does the organization spend its available money on people? On safety? Or are these ignored for other things?
- How people are measured: Is safety measured as tightly as production? What is measured tightly is what is important to management.
- How people are rewarded: Is there a larger reward for productivity than for safety? This states management's real priorities.
- Is teamwork mastered? Or is it "them versus us"? In safety is it "police versus policed"?
- What is the history—what are your traditions?
- Is your safety system in place to save lives or to comply with regulations?
- Are supervisors required to do safety tasks daily?
- Do big bosses wander around? Talk to people?
- Is using your brain allowed on the work floor?

- Have you downsized?
- Is the company profitable? Too much? Too little?

As you can see, an infinite number of things set the culture. We have listed only a few. It is more important to understand what the culture is than to understand why it is that way.

We have suggested that culture dictates what program elements will work and what elements will not work. Culture dictates final results, and what the accident record will be.

A fault tree suggested for analysis of the safety culture is shown in Figure 5-1.

For brainstorming the causes of a weak safety culture, you might again use a fishbone diagram (see Chapter 4).

MANAGEMENT CREDIBILITY

The perception of credibility is built over time and is based on a number of things: decisions, reward structures, measurements used, money spent or not spent, visibility, and so on. Measurable credibility is determined by what man-

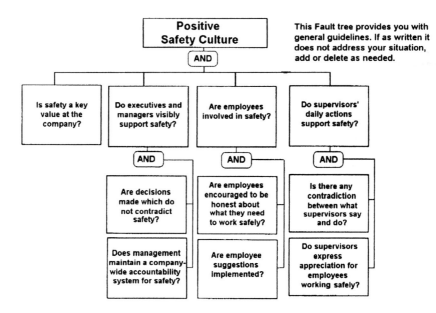

Figure 5-1. Positive safety culture fault tree.

agement has done in the past, not by what they have said. Management credibility in safety is earned over time and is based on the items just listed.

Top management participation is highly important in functioning safety programs. This is not exactly a new or exciting thought—safety people have spent considerable time thinking about and talking about how to get management's backing, how to enlist their support, or how to get their interest. For the most part, the safety people have little to show for their efforts.

Perhaps the real key to our past problems of getting management's "backing" is that the few times we have gotten top management's attention, we have not been sufficiently clear about exactly what we wanted them to do. We have given them our figures, which either show that we are doing a good job or that "they" (the line) are not doing their job. But we have not stated exactly what we want the executives to do differently from what has been done in the past. About the clearest thing we do ask is that they sign and issue a safety policy.

Our thinking on "managing the safety function" starts with a written safety policy, a definition, if you will, of management's desires concerning safety. Most executives agree that a policy is a fine thing to have, but few agree on what a policy is. "Policy" is often confused with "rules," "established practices," "procedures," and "precedents," not only in speech but also in action.

Too often in discussions regarding policy, we become bogged down trying to sort out policy from procedures, SOPs, and rules. The most important thing is probably whether management's interest is accurately communicated. If the principle "Safety is a line responsibility" is true, it is important not only that we as safety people believe it, but also that the line organization believe it. The line organization will believe that safety is its responsibility only when safety is definitely assigned to it by management.

Another principle is that "Management should direct the safety effort by setting goals and by planning, organizing, and controlling to achieve them." A safety policy is management's expression of direction to be followed. In almost all cases it is important that management's safety policy be in writing to ensure that there will be no confusion concerning direction and assignment of responsibility.

What is included in the safety policy may vary from company to company. No doubt most organizations will not write "pure" policy. They will include, whether intentionally or inadvertently, some procedures, some philosophy, and perhaps even some rules. This is perfectly all right—whatever serves the company best is what should be included.

We can list some of the things that should be included in most management policies on safety. At a minimum, the following concerns ought to be addressed in a safety policy:

- Management's intent: What does management want?
- The scope of activities covered: Does the policy pertain only to on-the-job

safety? Does it cover off-the-job safety also? Fleet safety? Public safety? Property damage? Fire? Product safety?
- Responsibilities: Who is to be responsible for what?
- Accountability: Where and how is it fixed?
- Safety staff assistance: If there is safety staff, how does it fit into the organization? What should it do?
- Safety committees: Will there be committees? What will they do? Why do they exist?
- Authority: Who has it, and how much?
- Standards: What rules will the company abide by?

Additional Executive Functions

Just signing policy clearly is not enough. The role of the executives must be spelled out in detail in the safety program. For instance, the responsible executives should:

- Sign and issue safety policy.
- Receive information regularly on who is and who is not performing in safety according to some predetermined criteria of performance.
- Initiate a positive or negative reward to immediate subordinates (middle managers).

On a day-to-day basis, this may be the only executive input needed in the program. At times, of course, policy decisions might be necessary, but they would be handled as any other mangement policy decision might be handled.

Management tends to delegate safety responsibility to the lowest possible level. Although this concept is excellent, there are some real hazards in its application. When all safety responsibility is delegated, upper levels of management are left with little to do, other than the traditional role of signing policy. Too often such a lack of involvement on a regular basis can be construed by the individuals in the organization as saying that safety is not important enough for executives to spend their time on it.

It is therefore extremely important that safety responsibilities not be delegated totally to the lowest possible level. All levels of management, including the top level, must retain some specific predetermined tasks that are visible and that can be seen by all workers in the organization as an indication of the importance of safety to the executives.

A fault tree that can be used to assess management's credibility is shown in Figure 5-2.

For brainstorming ways to improve your management's credibility, you might use a fishbone diagram.

The Management System to Build Culture   71

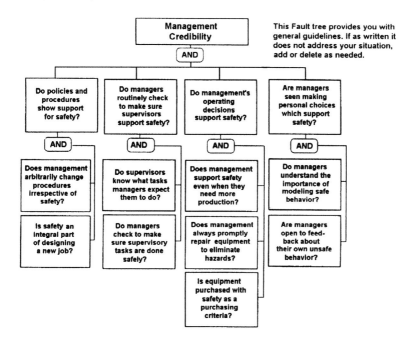

Figure 5-2. Management credibility fault tree.

SUPPORT FOR SAFETY

Safety results obviously require support and defined behaviors from the entire organization. Although safety support starts at the top, it is necessary throughout. This category is an indicator of employees' perception of how well the organization is working to create a safe work environment.

People in the organization will perceive that safety is supported when they see daily action aimed at preventing accidents. Daily action only happens when each person knows what to do (his or her role is defined) and is required to do it (is accountable).

Let us start by looking at one level, the first line supervisors. What drives their performance? At this level it is still simple. Performance is driven by the perception of what the next level supervisor wants done, these individuals' perception of how their supervisor will measure them, and their perception of how they will be rewarded for that performance. The research shows that the answers to the following questions dictate supervisory performance:

- What is the expected action?
- What is the expected reward?

- How are the two connected?
- What is the numbers game (how measured)?
- How will it affect me today and in the future?

The roles of people in the line hierarchy are as follows:

1. The role of the first line supervisor is to carry out some agreed-upon tasks to an acceptable level of performance.
2. The roles of middle and upper management are to:
   (a) Ensure subordinate performance.
   (b) Ensure the quality of that performance.
   (c) Personally engage in some agreed-upon tasks.
3. The role of the executive is to visibly demonstrate the priority of safety.
4. The role of the safety staff is to advise and assist each of the above people.

The supervisor's role as defined above is relatively singular and simple. It is defined as carrying out the agreed-upon tasks. What are those tasks? While it may depend upon the organization, the tasks might fall into these categories:

Traditional tasks	*Nontraditional tasks*
Inspect	Give positive strokes
Hold meetings	Ensure employee participation
Perform one-on-ones	Do worker safety analyses
Investigate accidents	Do force-field analyses
Do job safety analysis	Assess climate and priorities
Make observations	Perform crisis intervention
Enforce rules	
Keep records	

Middle managers also must have clearly defined roles, valid measures of performance, and rewards contingent on performance that are sufficient to get their attention. The performance at this level is critical to safety success. The middle managers (the persons to whom the first line supervisors report) are more important than the supervisors in achieving safety success because they either make the system run or allow it to fail. As stated earlier, the middle manager's role is:

- To ensure subordinate performance.
- To ensure the quality of that performance.
- To take actions that visually say that safety is important.

The Management System to Build Culture   73

There is probably more interest and commitment at the executive level today than we have ever seen before. Perhaps all of the new research is behind this interest and commitment. Perhaps the Bhopal, Chernobyl, and Challenger incidents explain some of it. The safety professional who does not take advantage of this interest is missing a major opportunity. In fact, executives may be interested today, but they typically do not have the foggiest idea of what to do to make safety happen. It is safety professional's job to spell out the role and to spell out the system. When all of management is actively involved in safety, the workers will believe it is a key value and that there is support for it.

A fault tree that can be used for assessing the organization's support for safety is shown in Figure 5-3.

For brainstorming ways to improve safety support, see Figure 5-4.

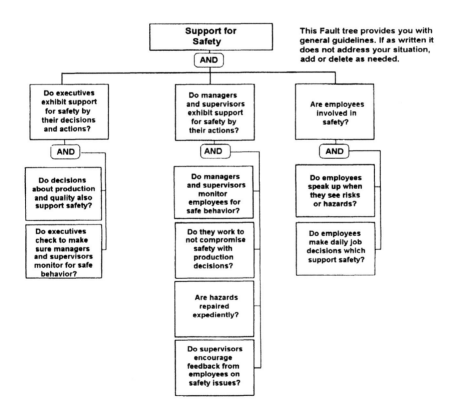

Figure 5-3.   Support for safety fault tree.

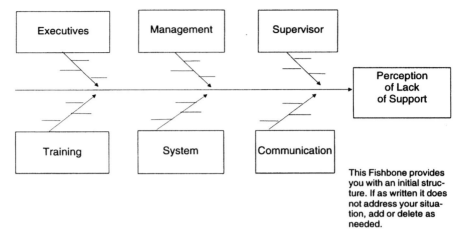

Figure 5-4. Fishbone diagram for improving safety support.

RECOGNITION FOR PERFORMANCE

Some organizations provide a means of recognizing good safety performance at all levels of the organization. However, recognition is almost always the lowest rated category in a perception survey, a reflection of the traditional management style. Only when there is a strong effort to ensure recognition does it occur. Many companies think of safety awards, contests, and trinkets as recognition. At best, these might be awards, but they have little to do with recognition. Recognition refers to whether or not workers feel as if they are being told that they are doing a good job. It refers to the one-on-one interpersonal recognition that is so often missing in a traditional management system.

Behavior modification (the use of positive reinforcement) is not new. The basic process involves systematically reinforcing positive behavior while at the same time either ignoring or using some negative reinforcements to eliminate unwanted behavior. There are two primary approaches used in behavior modification programs: one is an attempt to eliminate unwanted behavior that detracts from attaining an organizational goal, and the other is the learning of new responses. In safety, the primary objective is to eliminate unsafe acts. The second major goal of behavior modification is to create acceptable new responses to an environmental stimulus.

The basic concept of behavior modification is the systematic use of positive reinforcement. The result of using positive reinforcement is improved performance in the area to which the positive reinforcement is connected. The concept is based on the simple formula

The Management System to Build Culture

$$B = f(C)$$

which means that a person's behavior, B, is a function of the consequences of past behavior, C. If a person does something and immediately following the act something pleasurable happens, he or she will be more likely to repeat that act. If a person does something and immediately following the act something painful occurs, he or she will be less likely to repeat that act (or will be sure not to be caught in that act next time).

The use of positive reinforcement is gaining wider and wider acceptance in industry. It has helped to increase productivity and quality, improve labor relations, meet Equal Employment Opportunity (EEO) objectives, and reduce absenteeism.

A number of experimental programs have shown the potential of positive reinforcement in safety (see Figure 5-5). As the figure shows, the results are impressive, with reductions in unsafe behavior ranging from 8 to 350 percent with no negative results.

Safe behavior reinforcement is nothing more than recognizing people who do a good job at the time when they are doing it. Recognition is both the most underused management technique we have and the most powerful.

A fault tree that can be used to assess an organization's attempts at recognition is shown in Figure 5-6.

An individual's performance as a supervisor in providing recognition for safe behavior can be assessed by using Figure 4-3.

For brainstorming ways to improve recognition in your organization, use Figure 5-7.

ATTITUDES TOWARD SAFETY

The following are attitudes and behaviors generally observed in a good safety climate:

- Management considers safety to be important.
- Supervisors are seen as paying adequate attention to safety and getting support from management in that regard.
- Employees see management's effort to run a safe operation as "fair" and "effective."
- Employees support management's safety efforts and take an active interest in their unit's safety performance.
- Risk taking is discouraged at all levels of the organization.
- Supervisors have a positive attitude toward safety performance.

	RESEARCHER	YEAR	TYPE REWARD	MEASURES USED	RESULTS OBTAINED	PERCENT IMPROVE-MENT
1.	Komaki, Barwick & Scott	1978	Praise & Feedback 3–4 ×/wk	% Safe Behaviors	70% up to 95.8% 77.6% up to 99.3%	37% 29%
2.	Komaki, Heinzman & Lawson	1980	Feedback Daily	% Safe Behaviors	34.4% up to 68.4% 70.8% to 92.3%	98% 30%
3.	Krause	1984	Feedback	Unsafe Behaviors Lost Time Accidents Severity Rates		80% 39.3% 39.2%
4.	Sulzer-Azaroff & DeSantamaria	1980		Accident Freq.	15 to 0	100%
				Accident Freq. Hazards Hazards	45 to 33	27% 29% 88%
5.	Fellner & Sulzer-Azaroff	1984	Praise & Feedback	% Safe Behaviors	78% up to 86% 79% up to 85%	10% 8%
6.	Petersen	1983	Praise & Feedback	Safety Sampling	Two Railroad Divisions for Six Mos. with Control Groups	40% 49%
7.	Rhoton	1980	Praise & Feedback	Violation	1–4 per mo. to 0	100%
8.	Hopkins, Conrad & Smith		Feedback & Money	% Safe Behaviors Housekeeping Rate Housekeeping Rate	60% up to 100% 20 to 90 45 to 100	67% 350% 122%
9.	Chhokar & Wallin	1984	Feedback	% Safe Behavior	65% up to 81%	20%
10.	Uslan & Adelman	1977	Praise	Injury Frequency	81 50% % up to 95%	17%

Figure 5-5. Safe behavior reinforcement.

The Management System to Build Culture   77

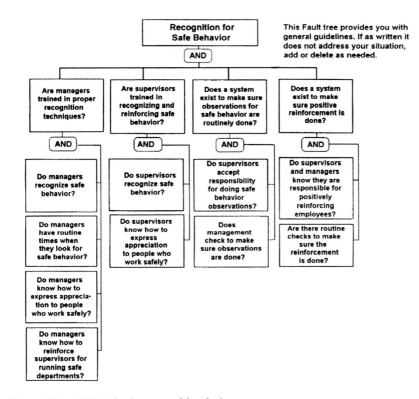

Figure 5-6.   Safe behavior recognition fault tree.

Attitudes are not addressed per se in a good climate; rather the attitude "score" is a reflection of the effectiveness of other parts of the safety system. The system must aim at behaviors; attitudes are the result. Good attitudes result from management systems that effectively influence behaviors of management, supervisors, and workers. Attitudes are not crisp and easily measurable, but, even so, we should attempt to recognize and assess employee attitudes toward safety.

Robert Mager's book *Developing an Attitude toward Learning* provides a lot of insight into attitude development, which can be applied directly to developing safety attitudes. Mager reports a study he made some years ago designed to determine students' attitudes toward different academic subjects and what formed the attitude in each case. To summarize his results, he states that a favorite subject area tends to become favorite because the person seems to do well at it, because the subject is associated with liked or admired friends, relatives, or instructors, and because the person is relatively comfortable when dealing with the subject. Conversely, a least-favored subject seems to become so because of a low aptitude for

78 THE AREAS TO ANALYZE

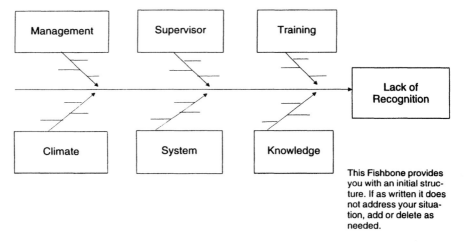

Figure 5-7. Fishbone diagram for improving recognition.

it, because it is associated with disliked individuals, and because dealing with the subject matter is associated with unpleasant conditions.

Things that help to mold an attitude toward a subject are:

- The conditions that surround the subject.
- The consequences of coming into contact with the subject.
- The way that others react toward the subject (modeling).

Conditions

When a person is in contact with a subject it should be dealt with under positive conditions, and, conversely, there should be no aversive conditions. If a subject that initially has no special significance is presented on several occasions to someone under an unpleasant condition, that subject may become a signal that triggers an avoidance response (to keep away). If a person is presented with a neutral subject and at the same time is experiencing pleasant conditions, that subject may become a signal for an approach response (to like it).

How can we use this information in our safety programming? First of all, if we do not already know what our employees consider to be positive and aversive conditions, we certainly ought to find out. To begin with, any form of punishment obviously is aversive; most forms of social interaction are positive; being "told" what to do is aversive. We should arrange to give our safety instruction under positive conditions, with as few aversive conditions as possible.

Consequences

Whenever contact with a subject is followed by positive consequences, the subject will tend to become a stimulus for approach responses. Conversely, whenever contact with the subject is followed by aversive consequences, the subject may become a stimulus for avoidance responses. This statement can be found in any freshman psychology text. It has been documented experimentally and practically, perhaps more than any other psychological principle.

Positives and Aversives

What in a work situation is aversive, and what is positive? Although it is not always possible to know whether an event is positive or aversive for a given individual, some conditions and consequences are universal enough to provide us some direction.

Mager suggests we define an aversive as any condition or consequence that causes a person to feel smaller or diminishes his or her importance. Here are some common aversives, adapted from Mager, that might apply to safety and safety training:

1. *Pain:* not too applicable to training, but this is learning by experience (the hard way) on the job.
2. *Fear and anxiety:* things that threaten various forms of unpleasantness, such as:
   - Telling the worker by word or deed that no accomplishment will bring success.
   - Telling the worker, "*You* won't understand this, but...."
   - Telling the worker, "It ought to be perfectly obvious...."
   - Threatening the exposure of "ignorance" by forcing the worker to do something embarrassing in front of the peer group.
   - Basing an attrition rate on an administrative flat rate rather than on worker performance. ("Half of you won't be here a month from now.")
   - Being unpredictable about the standard of acceptable performance.
3. *Frustration creators,* such as:
   - Presenting information in larger units, or at a faster pace, than the student can assimilate. (The more motivated the worker, the greater his or her frustration when efforts are blocked.)
   - Speaking too softly to be heard easily (blocking the worker's effort to come into contact with the subject).
   - Keeping secret the intent of the instruction or the way in which performance will be evaluated.

- Providing unreadable print—type too small or too ornate, or reading level too high.
- Providing obscure text or implying more profundity than actually exists, as in OSHA standards.
- Teaching one set of skills and then testing for another.
- Testing for skills other than those stated in announced objectives.
- Refusing to answer questions.
- Using test items with obscure meanings.
- Forcing all to proceed at the same pace, thus frustrating the slow and boring the quick.
- Calling a halt when the worker is absorbed with the instruction or attempting to complete a project.

4. *Humiliation and embarrassment,* for instance:
   - Publicly comparing a worker unfavorably to others.
   - Laughing at a worker's efforts.
   - Spotlighting a worker's weaknesses by bringing them to the attention of the group.
   - Belittling a worker's attempt to approach the subject with replies such as, "Stop trying to show off," or "You wouldn't understand the answer to that question."
   - Repeated failure.
   - Special classes for accident repeaters.

5. *Boredom,* caused by:
   - Presenting information in a monotone.
   - Insisting that the worker sit through repeated sessions covering the same topics.
   - Using impersonal, passive language.
   - Providing information in increments so small that they provide no challenge or require no effort.
   - Using only a single mode of representation (no variety).
   - Reading the safety rules aloud.

6. *Physical discomfort,* such as:
   - Allowing excessive noise or other distractions.
   - Insisting that the worker be physically passive for long periods of time.

Here are some positive conditions or consequences:

- Reinforcing or rewarding subject approach responses.
- Providing instruction in increments that will allow success most of the time.
- Eliciting learning responses in private rather than in public.
- Providing enough signposts that the worker always knows the rate of progress.
- Providing the worker with statements of your instructional objectives in an understandable mode.

- Detecting the worker's level of knowledge so as to avoid repetition (and the possibility of a very boring session).
- Providing feedback that is immediate and specific to the worker's response.
- Giving the worker some choice in selecting and sequencing subject matter, thus making positive involvement possible.
- Providing the worker with some control over the length of the instructional session.
- Relating new information to old, within the experience of the student.
- Treating the worker as a person rather than as a number.
- Using active rather than passive words during presentations.
- Making use of those variables known to be successful in attracting and holding people's attention, such as motion, color, contrast, variety, and personal reference.
- Making sure the worker can perform with ease, so that confidence can be developed.
- Allowing only those instructors who like and are enthusiastic about their subjects (and workers) to teach.
- Expressing genuine delight at seeing the worker.
- Expressing genuine delight at seeing the worker succeed.
- Providing instructional tasks that are relevant to your objectives.
- Using only those test items relevant to your objectives.
- Allowing workers to move about as freely as their physiology and their curiosity demand.

Modeling

Another way in which behavior is strongly influenced is through modeling (learning by imitation). The research on modeling tells us that if we want to maximize approach tendencies in workers, we must exhibit that behavior ourselves. In other words, we must behave the way we want our employees to behave. When we teach one thing and model something else, the teaching is less effective than if we practice what we preach. The message to safety management is obvious.

A suggested fault tree to assess attitudes toward safety is shown in Figure 5-8. For brainstorming about attitudes in the organization, see Figure 5-9.

STRESS

This category is concerned with the organization's propensity to stress claims. It looks at the environmental factors that lead to stress and stress claims, and at signs that indicate stress is present. The newest and potentially most costly

82   THE AREAS TO ANALYZE

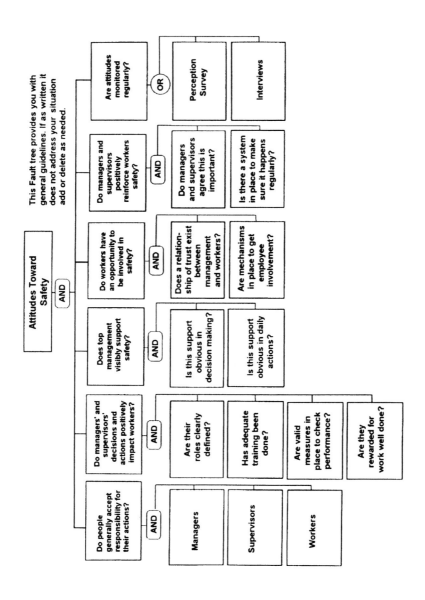

Figure 5-8. Fault tree for assessing attitudes toward safety.

The Management System to Build Culture

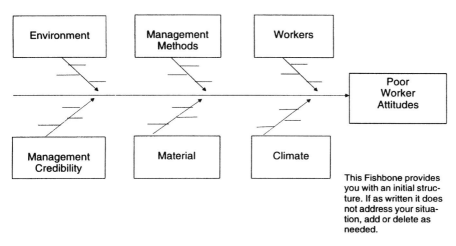

Figure 5-9. Fishbone diagram for improving worker attitudes.

segment of the subjective health-related illness problems in worker's compensation payout is due to psychological stress on the job. Although it is large enough to deserve much greater treatment than is possible in one section, here is an overview of the problem with some suggested control areas.

The problem is enormous. In 1900, our top killers were not stress-related illnesses; but by 1970, four of the top five killers were stress-related. In 1980, stress claims were minimal in the United States. By 1982, the problem had grown so that 11 percent of the worker's compensation health payout was due to stress-related problems.

By 1985, the problem represented the number one occupational health payout under worker's compensation. The Centers for Disease Control stated:[1]

> There is increasing evidence that an unsatisfactory work environment may contribute to psychological disorders.
> Studies show the factors contributing to unsatisfactory jobs include lack of control over working conditions, non-supportive supervisors or co-workers, limited job opportunities, role ambiguity or conflict, rotation-shift work, and machine-paced work.
> Mental stress can produce such illnesses as neuroses, anxiety, irritability, amnesia, headaches, and gastrointestinal symptoms.
> Average medical costs and indemnity payments in 1981–1982 for these forms of mental stress actually surpassed the average amounts for other occupational diseases.

The trends discussed by the Centers for Disease Control clearly indicate the future direction of stress-related illness. Without some legislative intervention, industry's problem could be massive. The primary reason for this is that a variety of diseases caused by stress could end up being compensable. Here is a partial list:

Coronary heart disease	Diabetes
Hypertension	Gout
Ulcers	Migraines
Colitis	Glaucoma
Anxiety	Epilepsy
Depression	Hemorrhoids
Allergies	Asthma
Arthritis	Acne
Cancer	Back pain

The reasons for the connection between stress and these illnesses can be better understood by examining the body's responses to stress. (Table 5-1).

Table 5-1. Bodily responses to stress—benefits and drawbacks

Natural Response	Original Benefit	Today's Drawback
Release of cortisone from the adrenal glands.	Protection from an allergic reaction to dust from a fight.	Cortisone destroying resistance to cancer, causing other illnesses.
Thyroid hormone increases in the bloodstream.	Body's metabolism speeded up, providing extra energy.	Shaky nerves, exhaustion, jumpiness, insomnia.
Release of endorphin from the hypothalamus.	A potent pain killer provided so one can't feel wounds incurred.	Aggravation of migraines, backaches, even pains of arthritis.
Reduction in sex hormones, testosterone, and progesterone.	Decreased fertility, a help in overcorwding and loneliness.	Anxieties and failures in sex, frustration, irritation.
Shutdown of the digestive tract.	Diversion of blood to muscles, giving more power for fight.	Dry mouth, bloating, diarrhea, discomfort, cramps.
Release of sugar into the blood, with insulin to metabolize it.	Quick energy supply to escape danger.	Diabetes, hypoglycemia.
Increase of cholesterol in blood from the liver.	Long distance energy provided to the muscles for fight or flight.	Coronary, heart problems.
Racing heartbeat.	More blood to muscles and lungs for a fight.	High blood pressure, strokes.
Increased air supply	Extra oxygen to the lungs for fight or flight.	A danger for smokers.
Thickening of the blood.	More capacity to carry oxygen; to fight infection from wounds.	Strokes, heart attacks.

The Management System to Build Culture

Our natural physiological responses to stress cause physical problems. Those responses (see Table 5-1) were completely appropriate in another age. For the early human faced with a stressful situation, the physiological responses prepared him or her for action to deal with the stress, either to fight or to take flight. Today the threat is usually not real, but holds only symbolic significance. Our lives usually are not in danger, only our egos. Physical action is not warranted and must be subdued, but for the body organs it is too late—what took only minutes to start will take hours to undo. The stress products that are flowing through the system will activate various organs until they are reabsorbed back into storage or gradually used by the body. And while this gradual process is taking place, the body organs suffer.

This fight or flight reaction has helped ensure our survival and continues to do so; no amount of relaxation training can ever diminish the intensity of this innate reflex. Stress is physical. The reaction permits a physical response to a physical threat. However, any threat, whether physical or symbolic, can bring about this response. Once the stimulation of the event penetrates the psychological defenses, the body prepares for action. Increases in hormonal secretion, cardiovascular activity, and energy supply signify a state of stress. It is a state of extreme readiness, enabling one to act as soon as the voluntary control centers decide the form of the action, which in our social situation is often no "action."

Natural Responses to Stress

The damaging effects of stress can be categorized as either (1) changes in the physiological processes that alter resistance to ideas, or (2) pathological changes, that is, organ system fatigue or malfunction, which result directly from prolonged overactivity of specific stress organs. There are a gastrointestinal system response, a brain response, a cardiovascular response, and a skin response—hence the connection to so many diseases.

By definition, stress is an arousal reaction to some stimulus, which can be an event, an object, or a person. It is characterized by heightened arousal of physiological and psychological processes. The stimulus that causes the arousal reaction is the stressor. Some typical stressors are:

Psychosocial
Adaptation   Overload
Frustration   Deprivation

Bioecological
Biorhythms   Noise
Nutrition

Personality

Self-perception   Anxiety
Behavioral patterns

Stressors

Is stress a safety problem? As already shown, some stress is clearly work-caused. All of the psychosocial stressors can occur at work. Anyone who has ever worked in an organization has experienced and is probably now experiencing the pressure to adapt to someone else's will and ideas; is experiencing some form of frustration (a block between oneself and one's goal); has been overloaded at some point, either physically or psychologically; or has felt deprived of filling some basic human needs on the job. Some of the bioecological and personality causes can also be job-related. Stress-related illnesses clearly can be job-caused, and there is little rationale to saying that they should not be compensable. The problem is that these same stressors also occur elsewhere: at home or in some social situations.

The more that people perceive themselves to be in control of a situation, the less severe their stress reaction is. This suggests that feeling helpless and sensing that one lacks sufficient power to change one's environment may be fundamental causes of distress. Thus, anything that enhances one's feeling of self-control is likely to reduce the severity of the stress reaction.

It is now generally accepted that experiencing stress is a significant part of performing a job for an employer or even for oneself (as an entrepreneur). Furthermore, empirical evidence indicates that excessive job stress is associated with negative health consequences. Despite general acceptance of the pervasiveness of stress and growing empirical evidence about stress's effects on health, only a limited amount of research rigorously evaluates the effectiveness of stress management programs within organizations.

The elements of a stress control program in an organization fall into two categories, those things the organization can do and those things that the organization can support the individual in doing:

Organizational strategies

Goal setting   Performance feedback systems
Participative decision making   Role specification
Job enrichment   Employee surveys
Work scheduling   Training programs
Culture change   Wellness programs

The Management System to Build Culture

Individual strategies	
Cognitive appraisal	Breathing
Restructuring	Progressive relaxation
Transcendental meditation (TM)	Biofeedback
Relaxation response	Exercise
Social support systems	Diet
Employee assistance programs	Cranial electrotherapy
Rehearsal	Stimulation

Stress Control Program Elements

These factors have to do with: improvement of the quality of goal setting at work; participation; job enrichment; improving the culture and the climate; performance feedback; specification of roles, responsibilities, and accountabilities (defuzzying things); surveying employee perceptions; training programs; wellness programs; and employee assistance programs.

Stress arises in mature adult human beings who feel that they are able to control

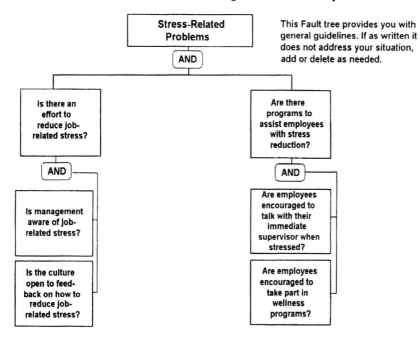

Figure 5-10. Fault tree for stress-related problems.

88   THE AREAS TO ANALYZE

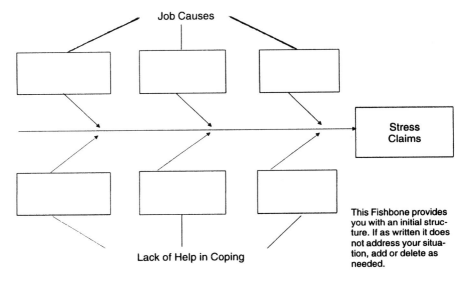

Figure 5-11. Fishbone diagram for organizational stress.

their own behavior but who find themselves in a work environment where they are under the control of someone or something else. The answer to stress problems, then, is for less control to be exerted over those individuals, or rather to shift that control from the organization to them.

A suggested fault tree to assess the organization's propensity to giving stress is shown in Figure 5-10.

For brainstorming about stress in the organization, see Figure 5-11.

Chapter 6

The Management System to Improve Managers' Skills

Managers, at all levels, usually are the key to safety success. We used to say that the foreman is the key man in safety. We now say it is the middle manager who controls supervisory performance. Either way, management action is crucial to results.

This chapter discusses the areas that ensure management performance.

SUPERVISORY TRAINING

The first question is, are supervisors perceived to be well-trained and able to handle problems related to safety? To answer this, some role definition is required—that is, defining the tasks or activities the supervisor should do. This can be accomplished authoritatively or participatively. Once they are defined, however, the supervisor must know how to do the tasks. This category provides an indication of whether the supervisors have been given sufficient training to carry out their defined role. The other essential elements of supervisory training are measurement and reward.

When training is done systematically, the organization uses this process:

1. It finds out where people are in their current knowledge and skills.
2. It finds out where people should be in terms of the behaviors required to perform the job safely.
3. It figures out a systematic way to provide the difference.

These three steps concentrate on defining training needs. Historically in safety training, the concentration has been on other areas, such as choice of training method or specific content. Learning theorists tell us to spend the bulk of our time on defining needs; if we do a good job of defining the need, everything else falls easily into place. The assessment of training needs involves a three-part analysis:

1. An organization analysis, to uncover company resources and objectives that relate to training.
2. A job analysis, to define corporate jobs and tasks.
3. A worker analysis, to explore the human dimensions of attitudes, skills, and knowledge as they relate to the company and the employee's job.

Organization analysis involves a study of the entire organization: its objectives, its resources, the allocation of these resources in meeting its objectives, and its total environment. These things largely determine the training philosophy of the entire organization.

Job analysis for training purposes involves a careful study of jobs within the organization in a further effort to define the specific content of training. It requires an orderly, systematic collection of data about each job, similar to job safety analysis procedures. The analysis might include the following questions:

- Is there obvious evidence of unsafe acts or poor procedures?
- Are there incidents on the part of individuals or groups that reveal poor personnel relationships, emotionally charged attitudes, frustrations, lack of understanding, or personal limitations? Do these situations imply training needs?
- Are there management requests for training of employees?
- Do interviews with supervisors, top management personnel, and employees reveal information about safety problems?
- Do group conferences (with interdepartmental groups and safety advisory committees), where organizational objectives, major operational problems, and plans for meeting objectives are discussed, reveal areas in which training could be of value?
- Have comparative studies of safe versus unsafe behavior been done to determine the basis for differentiating successful from unsuccessful performance?
- Are supervisors' reports on the safety performance of employees reviewed to determine training needs?
- Are accident records reviewed to determine the need for retraining?
- Is actual job performance observed?

Worker analysis focuses on individuals and their performance on the job as it relates to safety. A performance chart (see Figure 6-1) helps define individual

The Management System to Improve Managers' Skills

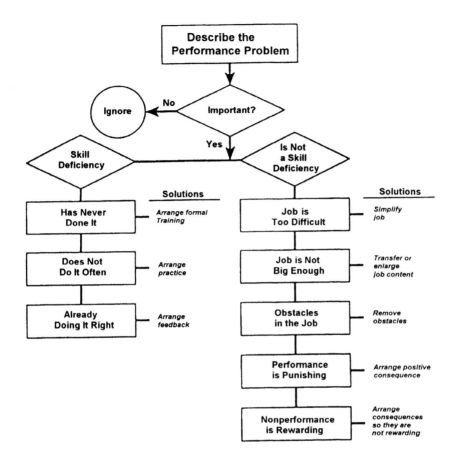

Figure 6-1. Performance model, for assessing an individual's performance as a supervisor in safety.

training needs. This performance analysis chart, which was devised by Dr. Phil Brereton of Illinois State University, helps us to tell whether the person does in fact have a training problem. Too often we seize upon training as the solution to almost any problem.

Finally, what is usually missing from the training design is evaluation. The evaluation of training is not simple. We try to determine what changes in skill, knowledge, and attitude have taken place as a result of training. Too often in safety training there simply is no measure of any kind to evaluate whether or not the training has met its objectives, or if it has accomplished anything; but there should be. Participants in the training process, regardless of organizational level, should show a measurable improvement in some skill or behavior, and management should insist that the training be evaluated in terms of improvement of the skill or measurement of behavior change. Thus, the performance model of Figure 6-1 (introduced in Chapter 4) can be used to assess an individual's performance as a supervisor in safety.

A suggested fault tree for assessing the effectiveness of supervisory training is shown in Figure 6-2.

QUALITY OF SUPERVISION

In a good system supervisors are perceived to be performing certain tasks related to accident prevention in a competent manner. Generally, these tasks include:

- Providing safety orientation for new employees.
- Recognizing and/or rewarding safe work behavior.
- Effectively handling employee concerns about personal matters.
- Discussing accidents and injuries with the employees involved.
- Discussing goals for safety performance with employees on a regular basis.
- Showing, by the emphasis they put on safety, that they are personally concerned.
- Involving their employees in all aspects of safety.

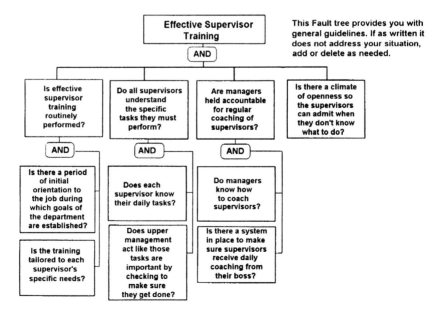

Figure 6-2. Fault tree to assess effectiveness of supervisory training.

Low positive perception in this category can mean several things: it can mean that supervisors turn over too rapidly to be effective; it can mean they simply do not know what to do to be effective; most likely, however, it indicates a lack of accountability—supervisors are not required to perform in safety because of invalid performance measures or a low reward structure.

Supervisory Performance Model

The first and most important principle of modern safety management is that "the key to line safety performance is management procedures that fix accountability." This is simply a restatement of a key management principle. Another way to state this is, "What gets measured and rewarded gets done."

Figure 6-3 shows a supervisory performance model (for safety or any other performance). The model suggests that whether or not your supervisors do anything to satisfy their safety responsibility is dependent on:

1. Their knowing what to do (task definition).
2. Their knowing how to do it (training).
3. Their knowing that the boss is measuring whether or not they do it (measurement).
4. There being a reward for doing it (reward).

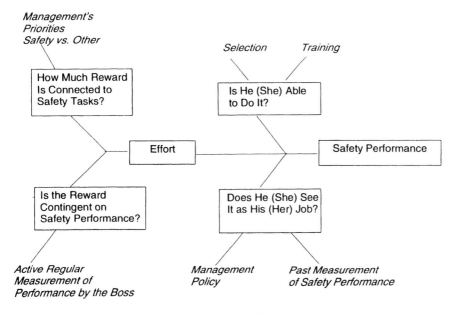

Figure 6-3. Supervisory safety performance model.

What drives performance is employees' perceptions of what the boss wants done, their perceptions of how the boss will measure them, and their perceptions of how they will be rewarded for that performance. To restate what the research shows, these issues dictate supervisory performance:

- What is the expected action?
- What is the expected reward?
- How are the two connected?
- What is the numbers game (how is it measured)?
- How will it affect me today and in the future?

Accountability Systems

Any accountability system that defines, validly measures, and adequately rewards will work.

A suggested fault tree to assess the effectiveness of supervisors is shown in Figure 6-4.

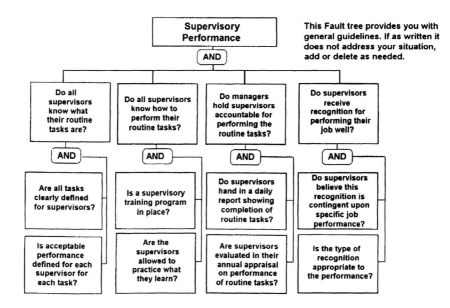

Figure 6-4. Fault tree to assess effectiveness of supervisors.

The Management System to Improve Managers' Skills

The individual's performance as a supervisor in safety can be assessed with the model shown in Figure 6-1.

For brainstorming on the effectiveness of supervision, the fishbone in Figure 6-5 is provided.

In addition these SPC tools can be used:

- *Paretos,* to zero in and assess the departmental supervisor's success, using perception survey results, behavior sampling results, interviews, audits (if valid), and even some accident statistics.
- *Control charts,* on behavior sampling results or statistics.
- *Flow charts,* to determine how the organization prepares individuals to become effective supervisors.
- *Scatter diagrams,* to determine what variables correlate with results.

For an explanation of these tools see Chapter 13.

GOALS FOR SAFETY PERFORMANCE

A good safety system has a method of setting goals for safety performance that involves all levels of the organization. It also communicates these goals to all concerned employees on a regular basis, together with information on performance related to these goals. Goals here involve more than just statistics. Goals are also (and more important) set for activities and process improvement. Goals

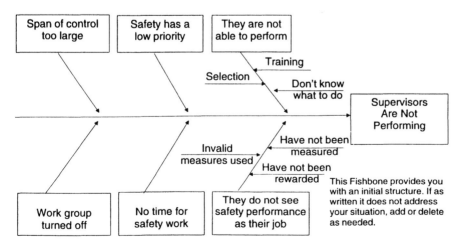

Figure 6-5. Fishbone diagram for use when supervisors are not performing.

should be set jointly between workers and management for what specifically has to be done to improve the safety process.

At most levels of the organization, it is more important to concentrate on activity goals than results goals. Activity goal setting ensures continuous improvement of the safety process—the results will take care of themselves. Goals seem to fall into specific categories, and in most situations we are setting goals of more than one type. The categories are:

1. *Routine goals.* These are the objectives that relate to the regular, repetitive, commonplace activities that are the normal, basic duties and requirements of the job. For instance, a routine goal for department supervisors might relate to the number of inspections they make.
2. *Project goals.* These goals are for new or special projects that might be undertaken by a manager in a coming time period. They also could relate to problem-solving or emergency-action activities that are a part of the supervisory job.
3. *Creative goals.* Although less commonly used, these goals relate to innovation or improvements in the department or in the management system.
4. *Personal goals.* Even less common, but sometimes used, are goals that relate to personal efforts that employees might make to supplement their job skills with new skills to increase their personal effectiveness.

All of the above are certainly feasible, allowable types of goals in safety management. In most cases, goals will be set in more than one of the above categories. Perhaps a number of goals will be set in the routine activities of a supervisor's job, but objectives also will often be set in the other three categories.

Routine Goals

The most common, and perhaps the most important, are the routine goals. These are the goals we usually think about when we set up a system. Most discussion and actual goal setting evolve around the determination of routine goals.

Routine goals do not necessarily include set goals relating to the number of accidents. That type of goal is only one of many types of routine goals, and is probably one of the least effective types. Numbers or costs of accidents are results goals. Probably at the lower levels of the organization, it is much better to set goals in terms of activities, the specific performances that the supervisor will engage in.

Results objectives generally relate to what is to be accomplished and are usually thought of in terms of failure measures: number of accidents, frequency rates, severity rates, cost of accidents, and so on. Activities or performance measures usually relate to the specific performances desired, such as number of inspections to be made, number of orientation talks to be given to new employees, and so forth. Both results and performance measures are considered to be routine goals.

In terms of motivating supervisors and getting results, it is always better to put the primary emphasis on performance goals.

Project Goals

Project goals can be used a great deal. They involve specific safety problems that can be worked on and solved by a supervisor; for instance, to study and learn how to effectively guard a certain machine or process. Project goals can be individual (something that just the supervisor will accomplish), or they can be group goals (the supervisor will solve a specific problem through the employees). Here again, the field is wide open; almost anything can be used as a project goal.

Creative Goals

Creative goals are a less common type, but they could be used more often and quite effectively. A creative goal might refer to the kinds of things a supervisor could do to improve the safety system generally. Many good improvements in safety programming generally can and do come from the line organization. A good creative goal might be that a specific supervisor, or many supervisors, will submit within the next period specific suggestions for the improvement of the corporate safety program. A creative goal for a supervisor might be to develop better, newer ways to tap the resources of the employees in safety or to obtain employee participation in safety.

Here too there is literally no end to the things that might be done that would fall under the heading of creative goals.

Personal Goals

Personal goals in safety probably will be a part of most objectives. A personal goal may be one of many objectives that are set. A personal goal might relate to a supervisor's self-development in safety: taking a special course in safety or in one aspect of safety for the next period. It could relate to a supervisor's self-study and the attainment of some new knowledge in an area related to safety and to the job. Personal goals could also relate to a supervisor's own skills and their improvement. Perhaps the supervisor needs to be a better counselor to the employees. An unlimited number of personal goals might be set.

Results Goals

Results goals fall into two broad categories: after-the-fact goals and before-the-fact goals. After-the-fact (results) measures are those that are based on incidents (accidents, near-misses, injuries, costs, etc.). Although they are important, we do not rely solely on them. Before-the-fact goals are better, such as behavior sampling.

Criteria for Setting "Good" Objectives to Achieve a Goal

If the objective-setting process is to work in our organizations, the objectives that are set have to be "good." Perhaps we should concentrate on the setting of objectives and attempt to identify what constitutes a "good" objective as opposed to a "poor" objective, and to see what criteria we might be able to establish. First, we can set criteria aimed at how the objective is devised and written, and then we can set some further criteria aimed at the relationship between the participants (the subordinate and superior managers) in the objective-setting process.

A good objective will satisfy four criteria pertaining to its direction, its individuality, its measurability, and its realism in the situation that exists. These four criteria might be summarized by the acronym RIMER. The letters stand for the fact that a good objective is *R*ifled directly at specific performances, is *I*ndividual, (i.e., under the direct control of the subordinate), is *ME*asurable, and is *R*ealistic. Let us look at each of these criteria:

1. *Rifling of objectives.* This merely refers to the fact that a good objective aims at a particular and specific area of performance, as opposed to rather general, or "shotgun," objectives that might aim at general, nonspecific performance areas. For instance, an objective such as "to be a better supervisor next month" is so general that it is useless. A better example is: "The supervisor will interact with each of his or her employees two times every week."
2. *Individuality of objectives.* This refers to the objective's being very specific to the subordinate for whom it is set. An individual objective requires results and performance that the individual subordinate has enough control to perform on his or her own. An objective for the staff safety director to improve the record in Department A is an objective that is not specific to the safety director, for to attain the objective that safety director is dependent on the performance of the supervisor of Department A.
3. *Measurability of objectives.* If we are going to measure the performance of a manager toward reaching a goal, we must be able in some way to

measure that performance. Thus, the goal must be measurable. For instance, a goal of "to become a better listener in the next six months" is unmeasurable and, hence, a poor objective. However, "There will be no instances in the next six months where employees state: 'He never listens' " is a measurable goal.

4. *Realism of objectives.* Objectives must be based on facts and analysis of past performances, or the data, so that they are not merely dreams and wishes. To be good, objectives must be attainable and realizable. A personal goal "to be a millionaire next year" is simply not a realistic objective for most of us.

Thus, we say that a good objective is a RIMER. A good objective meets the four criteria of being Rifled, Individual, MEasurable, and Realistic.

All of these criteria speak to the objective itself. Other criteria can be established about how the objective will be reached. In this area, we might say that a good objective would meet three additional criteria, LUM:

1. It would be *L*eveled, that is, aimed directly at the organizational level of the person it is for. A goal that speaks to the results of a supervisor's department in a specific area is at the supervisor's level if he or she is the direct supervisor of that department. An objective of "helping the corporation attain the desired frequency rate" is not an objective that speaks to a supervisor's level.
2. It must be *U*nderstood by both sides, the superior and the subordinate. If it is not clearly grasped by both sides, it is a poor objective. Misunderstanding is one of the biggest reasons for unattained objectives.
3. It must be *M*utually agreed upon: neither party in the process may impose his or her will on the other. Without mutual agreement, the objective becomes poor.

Thus, in addition to our criteria that an objective must be a RIMER, we also say that a good objective is a LUM: at the proper organizational level, fully understood by both sides, and mutually agreed upon.

Objectives go wrong for many possible reasons, but the most likely seem to be summarized by the acronym POTABLE, which stands for:

1. It was a *PO*or objective to start with because it did not meet the RIMER or LUM criteria.
2. There was a lack of needed *T*raining or skill.
3. There was an *A*ttitude change.
4. There were *B*arriers in the *L*ine or *E*xecutive ranks.

100 THE AREAS TO ANALYZE

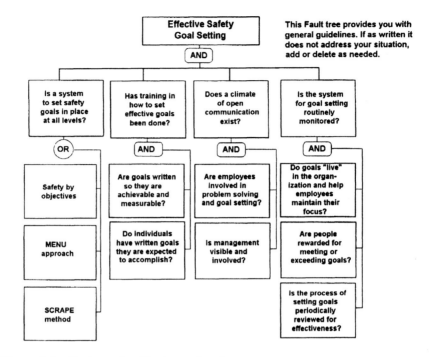

Figure 6-6. Fault tree for effective goal setting.

A suggested fault tree to assess the effectiveness of goal setting is shown in Figure 6-6.

Chapter *7*

The Management System to Improve Employee Skills

Management has the direct and sole responsibility to ensure that each employee has the skills needed to perform the job. This is true whether we are talking old style traditional management, today's management without supervisors at times, using temporary employees, or whatever is going on at one's company. People must know how to do what they are asked to do. This chapter discusses the areas this involves.

EMPLOYEE TRAINING

A good system provides training for employees in the aspects of their jobs that involve safety. The employees will think that they have received adequate training and that they understand how to work safely.

Employees' performance in safety depends upon their ability and motivation. Training enables them to perform safely; so training is essential. Training is (rightly or wrongly) by far the most frequently used method to change behavior in industrial safety.

We hire all types of people. Some are susceptible to accidents; others are not. Perhaps some are actually accident-"prone." Some are knowledgeable about safety from previous jobs. Some are eager to work safely, whereas others are indifferent to safety. Thus we have a total work force of people who may or may not have to be trained in safety. Where do we go from here? The logical process would be to attempt to answer these questions:

- Where are we going?
- How will we get there?
- How will we know we have arrived?

Management tells us, through its policy statements, where we want to be and what each person's function is once we get there. We then must find out how able the employees are to fulfill these duties and responsibilities. "Can they do it?" Defining needs is the key point in the process; it provides the objective and sets the criteria for measurement. We would expect any function this vital to the success of the training to be widely used in industry.

One study found that a careful and systematic investigation was conducted in only one out of ten companies. Very little training effort is based on any systematic appraisal of training and development needs. Instead, as reported by Dr. Donald Newport, we choose or prescribe training according to past rituals, current politics, and faddish misconceptions. He describes the most popular approaches in four industries. First, there is the "bandwagon" approach where a company tries to keep up with the incorporated Joneses. Second, there is the "smorgasbord" approach, which allows favorite employees to pick and choose which courses offer the newest buzz-words. Third is the "crisis" approach, which uses hindsight as its methodology; it is not needed until the crisis arrives, and corrective rather than preventive action is required. Finally, the "excursion" principle is used to measure course value—a program held 1,000 miles from home is twice as good as one only 500 miles away.

The assessment of training needs requires a three-part analysis (as described in Chapter 6):

1. Organization analysis—determining where within the organization training emphasis can and should be placed.
2. Job analysis—determining what should be the content of training in terms of what an employee must do to perform a task, job, or assignment in an effective way.
3. Worker analysis—determining what skills, knowledge, or attitudes an individual employee must develop if he or she is to perform the tasks that constitute his or her job in the organization.

Organization analysis involves a study of the entire organization, to obtain a clear understanding of both short- and long-term goals. What is the company trying to achieve in safety in general and specifically by department? Also needed is an inventory of the company's attempts to meet goals through its human and physical resources. Another important step is an analysis of the climate of the organization.

The climate of an organization is, in essence, a reflection of its members' attitudes toward various aspects of work, supervision, company procedures, goals and objectives, and membership in the organization. These attitudes are learned;

they are a product of the members' experiences both inside and outside the work environment. A training program may be designed to effect certain changes in the organizational climate. For instance, our safety training certainly hopes to influence the employees' attitudes toward safety.

Job analysis for training purposes involves a careful study of jobs within an organization in a further effort to define the specific content of training. It requires an orderly, systematic collection of data about the job, which we are familiar with through our job safety analysis procedures. The following methods are also available for job analysis:

- Observations: Is there obvious evidence of unsafe acts or poor methods? Are there occasions when individuals or groups reveal poor personnel relationships, emotionally charged attitudes, frustrations, lack of understanding, or personal limitations? Do these situations imply training needs?
- Interviews with supervisors and top management personnel to accumulate information about safety problems, as well as interviews with employees concerning safety.
- Group conferences with interdepartmental groups and safety advisory committees to discuss organizational objectives, major operational problems, plans for meeting objectives, and areas in which training could be of value.
- Comparative studies of safe versus unsafe employees to underline the bases for differentiating successful from unsuccessful performance.
- Questionnaire surveys.
- Tests or examinations of safety knowledge of current employees; analyses of safety sampling.
- Supervisors' reports on the safety performance of employees.
- Accident records.
- Actually performing the job.

Worker analysis focuses on the individual and job performance as it relates to safety. Use of a performance chart (see Figure 6-1) can help with this analysis.

How to Train

There are more empirical data here than perhaps anywhere else in psychology. Here are some of the "knowns."

Motivation and Learning

Learning theorists generally agree that an individual will learn most efficiently when motivated toward some goal that is attainable by learning the subject matter presented. It is necessary for the goal to be desired, and the learning behavior must appear to relate directly to achieving that goal. If the training does not seem to relate directly to the goal, other kinds of behavior that, to the learner, appear to be relevant to the goal will be tried.

In conducting a safety training course, for example, some people may think that they have more important production problems to worry about, and will spend their training time thinking about them and complaining about being taken away from the job to learn a lot of nonsense. People's behavior is oriented toward relevant goals, whether these goals are safety, increased recognition, production, or simply socialization. People attempt to achieve those goals that are salient at the moment, regardless of the trainer's intent. Therefore, making sure that employees understand the direct relationship between the training and the goal is critical.

Reinforcement and Learning

Positive rewards for certain behavior increase the probability that the behavior will occur again, whereas negative rewards decrease the probability. Whether or not an event is reinforcing will depend on the perceptions of the individual who is learning. What one person regards as a rewarding experience may be regarded by another as neutral or nonrewarding or even punishing. In general, however, there are various classes of reinforcers: food, status, recognition, money, companionship. These are reinforcing to almost everyone at one time or another.

Practice and Learning

An individual learns what is practiced. In practicing a skill, those behaviors that are performed and reinforced are acquired and maintained. Without practice, learned skills are lost quickly. In safety training, then, follow-up and practice are as important as the initial learning. Also, spaced practice (a little at a time) seems to be more effective than massed practice. This seems to be true both for learning rates and for retention. It is better to have a number of short sessions than one long one.

Feedback and Learning

Most experts state that giving a trainee feedback is essential for good training. It is difficult for the trainee to improve his or her performance unless given specific knowledge about that performance. If trainees are not performing correctly, they need to know the nature of their errors. How can they be corrected? This is essential. Some theorists strongly emphasize immediate reinforcement to each bit learned.

Meaningfulness

In general, meaningful material is learned and remembered better than material that is not meaningful; so the material must be made as useful as possible to the trainee. The concept of meaningfulness has implications for the way in which material is presented to the trainee. The trainer must try to think in the trainee's terms, to put the material across with familiar examples and language. It is important to supply as many word associations for new ideas and concepts as possible so that they become more meaningful.

Climate and Learning

The classroom environment makes a difference. Researchers have concluded that to encourage high rates of achievement in highly technical subjects, the environment must be challenging. To encourage achievement in nontechnical areas, classes should be socially cohesive and satisfying.

These, then, are a few insights into the behavior of people involved in learning. Whichever methods are chosen, they should take into account the things we know from the research.

Evaluation of Training

This topic was discussed in some detail in Chapter 6. A suggested fault tree is shown in Figure 7-1.
  The individual performance model is that shown in Figure 6-1. The fishbone for brainstorming is shown in Figure 7-2.

NEW EMPLOYEES

A good safety system is supported by employment practices that reduce the chances that new employees will have accidents. Some of these include:

- Stressing safety policies in employment interviews.
- Hiring only those people fit to perform the job's duties.
- Orientation and training that stress the safety aspects of the job.
- Assignment to work with safe employees.

As the effectiveness of the selection process is severely limited by law, the

THE AREAS TO ANALYZE

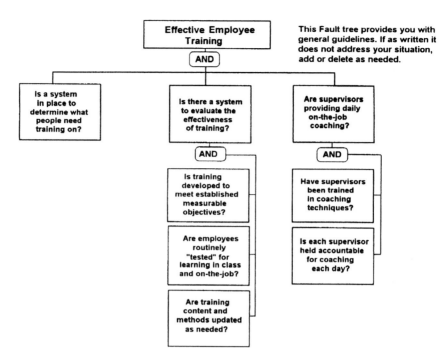

Figure 7-1. Fault tree for effective employee training.

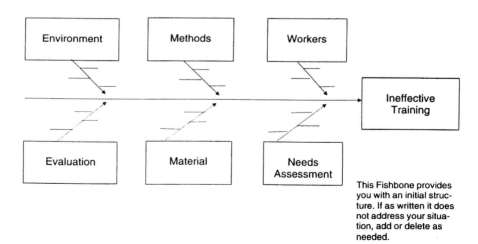

Figure 7-2. Fishbone diagram for ineffective training.

emphasis today must be on what occurs once an employee is on the job and on efforts to ensure his or her safety now as well as his or her future safe behavior.

Perhaps the logical starting point is to select people who will work safely—to screen out the "unsafe" workers and hire the "safe" ones. Obviously this is easier said than done. In the early years of safety, selection was one of the biggest tools. We believed we could select those applicants who would be safe workers and screen out the unsafe. It is not that simple, for the following reasons: People used to believe that certain applicants were "accident-prone," that is, with permanent susceptibility to accidents. But three problems emerged:

1. Theorists began to question the concept.
2. No one ever figured out how to test for "proneness."
3. Such testing probably would be illegal.

Current theorists seem ambivalent about the concept. In the United States, researchers tend to reject the proneness concept, but in other areas it is still believed. In all probability the other countries are on the right track; for research has shown a link between personality types and risk taking (which leads to accidents), and between what people have had to cope with and having accidents (a temporary susceptibility).

Research has shown that accident proneness is real but only in a tiny percentage of people, where it is probably a system of maladjustment. We probably cannot afford to seek out those people.

The typical selection process is shown in Figure 7-3. As the figure shows, there are two basic sources of information about an applicant: biographical data and test results.

The value and the validity of most of these selection devices have been fairly well scrutinized by researchers in the field of occupational psychology, and their research leads to these conclusions:

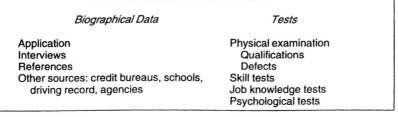

Figure 7-3. Employee selection.

- Job knowledge and skill tests can be administered effectively, provided that job criteria can be established in these areas.
- Similarly, physical exams can be helpful, provided that physical job criteria are available.
- Interviews are generally invalid, not because the applicant is lying but because the interviewer's biases and stereotypes invalidate them.
- Checks with previous employers (if we have established that there was a good working relationship) will provide an applicant's past history (assuming this is indicative of his or her future, which it generally is not; also the information probably cannot be proved).
- In addition, there are the further selection constraints of law. Most tests are illegal in the United States under the Equal Employment Opportunity laws, unless they have been statistically validated to job success at one's own company.

Training

Once selected, the new employee must receive orientation and job training. These topics are discussed in detail in Chapter 6, under "Supervisory Training."

Supervision

A new employee must be supervised differently from an experienced employee until fully acquainted with and skilled in the job. This requires much one-on-one attention (see discussion of the categories quality of supervision and communication in Chapter 6 and this chapter).

A suggested fault tree is shown in Figure 7-4.

A fishbone that can be used for brainstorming in groups is shown in Figure 7-5.

COMMUNICATIONS

In safety matters, it is essential for communication between employees and supervisors to go in both directions.

1. Employees regularly receive information on:
    (a) Cost, frequency, and type of accidents.

The Management System to Improve Employee Skills 109

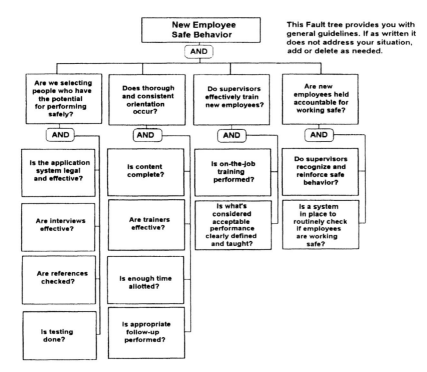

Figure 7-4. Fault tree for new employee safe behavior.

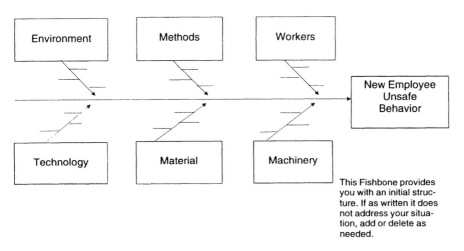

Figure 7-5. Fishbone diagram for new employee unsafe behavior.

110 THE AREAS TO ANALYZE

    (b) Hazards of the operation they perform and safe methods of operations.
    (c) Goals for safety performance and unit standings.
    (d) Safety rules.
2. Supervisors are perceived to be knowledgeable on safety matters.
3. Supervisors and management regularly revise information on what is working and what is not; what problems exist; what the solutions are to those problems; which procedures are needed; which procedures are not needed; etc.

Most of the communication process results from what happens every hour of every day, in the one-on-one ongoing regular relationship between the worker, his or her supervisor, and management. Formal communication is much less important than informal interaction. A simple definition, which suggests a simple model, is that communication is a process by which senders and receivers of messages interact in a given social context. For our purposes the senders are management (including staff safety and line supervision), the receivers are workers, and the social context is the industrial organization. A simplified version of the model might look something like that shown in Figure 7-6.

Management Variables

Three factors influence the effectiveness of communication: its credibility, its attractiveness, and its power (or control). Employees have clear-cut ideas about

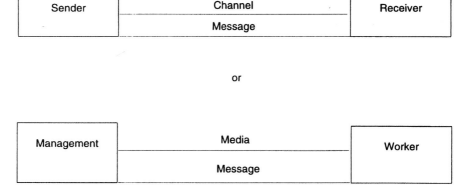

Figure 7-6.   Communication model.

their company's safety activities. The methods, the goals, and the amount of vigor and sincerity they reflect, as well as the way a worker views company safety efforts, strongly influence the employee's behavior on the job and ability to learn from and respond to safety media. In short, the effectiveness of the safety message presented by safety media (posters, films, booklets) is directly dependent on employee perception of management's interest.

Employee concerns to be considered in safety messages are: (1) What is management's credibility? Are members of management really as interested in my safety as they are trying to tell me? How have management demonstrated that interest in the past? (2) Is my supervisor really "like me" enough to be believable or so "different" as not to be trusted? (3) Is my supervisor's power over me real? Must I really comply with the safety rules?

What happens when workers receive conflicting messages from two different sources: management and their work group? Almost invariably, we find that people are more influenced by their peers than by their superiors.

One-on-one communication by a supervisor is absolutely crucial to safety success; and because a supervisor's credibility is essential to the success of the one-on-one communication process, supervisors must seem credible to their subordinates.

Supervisory credibility is a function of the following behaviors:

1. Delegating responsibility in decision-making to subordinates.
2. Asking subordinates' opinions concerning upcoming decisions.
3. Giving subordinates opportunities to offer additional ideas or information over and above what the supervisor has asked for.
4. Giving prompt answers to questions and suggestions.
5. Making sure that subordinates find it easy to get help with their problems and complaints.
6. Being aware of and responsive to subordinates' feelings and needs.
7. Being "frank" and "open" with subordinates.
8. Being supportive of subordinates concerning the complaints to upper management.
9. Expressing a sincere concern for the welfare of subordinates by:
    (a) Maintaining reciprocal relationships by exchanging ideas with subordinates.
    (b) Showing interest in the personal lives of subordinates (but being careful to avoid any appearance of sexual harassment).
    (c) Being helpful when help is needed.
    (d) Being concerned about subordinates' getting ahead in the organization.
    (e) Being supportive with upper management.
    (f) Complimenting subordinates.

In the one-on-one relationship between a supervisor and a subordinate, the closer it is to a level relationship (that is, a relationship of equality rather than superior–subordinate), the better off both parties will be. You can have that level relationship one-to-one; you cannot have it in a group or meeting format. Also, motivation is defined by a person's attempting to satisfy his or her current needs. A supervisor can only find out what those needs are in a one-to-one format—not in a group or meeting setting; and the power of the informal group is the largest single determinant of individual worker behavior. Management can only break through that power one-on-one; for in a meeting or group format, that power is only strengthened.

Message Variables

Attitude-change research has focused more on message variables than on any of the other communication variables. It is known that the general skill of the sender has not proved to be a very powerful determinant of persuasive effectiveness. Although well-organized messages have been found to be more effective than poorly organized ones in affecting comprehension, they have no special effect on opinion or change. A second area of interest is comparison of dynamic with subdued styles of presentation. A dynamic presentation is less effective in producing attitude change and more likely to be labeled "propaganda" than is a conversational delivery. Very little research has been done on humor, but it has been found that the use of humor in a speech on a serious topic makes it neither more persuasive nor more interesting.

Receiver Variables

Another factor that communication researchers have studied is the receiver, in our case the worker. What are some of the variables here? The following research findings might give us some direction:

1. Learning is enhanced with participation. For instance, calling on a subject to improvise a speech produces more attitude change than having the person passively read an already prepared speech.
2. The relationship between individual susceptibility and persuasion is very complex. For instance, there are relationships between sex and influenceability, age and influenceability, background and influenceability, and education and influenceability; but the research results are confusing and, at times, conflicting. While we know they exist, right now it does not appear that we can successfully use these relationships.

The Management System to Improve Employee Skills 113

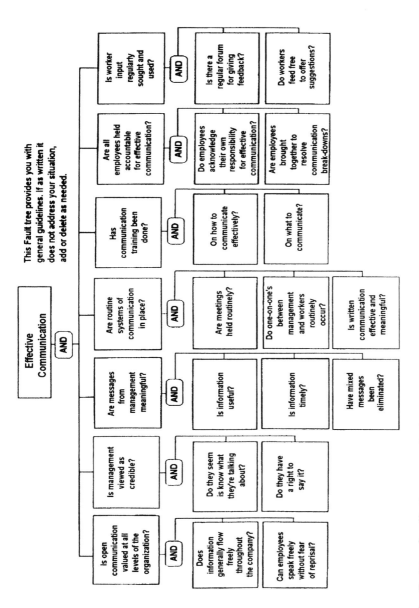

Figure 7-7. Fault tree for effective communication.

114   THE AREAS TO ANALYZE

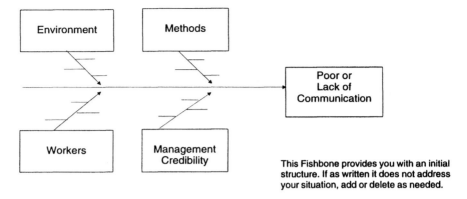

Figure 7-8. Fishbone diagram for poor or absent communication.

   3. The amount of attitude and behavior change depends to a great extent on the group and group norms with which the person is involved.

A suggested fault tree is shown in Figure 7-7.

A standard fishbone can be used, as shown in Figure 7-8.

Chapter 8

The Management System to Improve Worker Behavior

Once employees are on the job, it is incumbent upon management to do all they can to ensure safe behaviors. Some relevant factors are discussed below.

SAFETY CONTACTS

Contacts include safety meetings, one-on-one discussions, and other means used by supervisors and management to communicate. Although safety meetings are traditional, they may be less effective than one-on-ones. This category measures employee perception of contact effectiveness.

Safety contacts can be one-on-one discussions or occur in a meeting format. Success in each case depends upon the skill of the team leader, or supervisor. Here are some guidelines to effective contacts:

1. Before communicating, a person should analyze his or her problem in as much detail as possible to determine exactly what he or she wishes to communicate.
2. The purpose of the communication should be defined. Specify your intentions; then define your aim.
3. The physical and human environment should be considered, that is, timing, location, social setting, and previous experience.
4. Communication is not exclusively verbal. Consult with others if consultation is thought to be necessary.

5. Objectivity is not necessarily a criterion of good communication in every circumstance; sometimes two-sided messages are useful.
6. Try not to influence the persons with whom you are communicating; try to see things from their point of view. Remember that different people have different perceptions.
7. Assess the effectiveness of the communication if this is possible; this is usually done by encouraging feedback.
8. Choose carefully the type of communication process best-suited to your purpose.
9. Many executives seem to believe that it is possible to manage by exclusively vertical forms of communication, but research reveals that a great deal of effective communication must be lateral.

Research on "upward communication" reveals several conditions that appear to encourage upward communication:

1. *Frankness with management.* Establish genuine two-way communications between all levels of management. When critical discussion is choked off at higher levels of the company, it ceases to flow at lower levels.
2. *Supervisor accessibility.* Develop an awareness among managers that the keys to better listening are accessibility and responsiveness. Employees do not want to be heard all the time. But when they do have a problem, they need assurance that their supervisor will listen and act.
3. *Welcoming the new and different.* Tolerate all kinds of ideas—those that are foreign, silly, or hostile as well as those that management considers constructive, that is, those that it is willing to accept. Looking with disfavor on employees for thinking differently leads to closed minds.
4. *Visible benefits.* Visibly reward those who have creative new ideas. This is the strongest encouragement management can give.
5. *Acceptance of criticism.* Regard criticism as healthy and normal and lack of criticism as dangerous and undesirable, an indication that employees have given up trying to get through to management.
6. *Sensitivity to the employee.* Be willing to wrestle with the problem of interpreting what an employee is really trying to say. An employee's gripe about working conditions may mask a belief that the supervisor does not appreciate his or her job performance.

The importance of the supervisor's credibility was discussed in Chapter 7. Supervisory credibility is a function of the following behaviors, among others:

1. Delegating responsibility in decision-making to subordinates.
2. Asking subordinates' opinions about upcoming decisions.

3. Giving subordinates opportunities to supply additional ideas or information over and above what the supervisor has asked for.
4. Giving prompt answers to questions and suggestions.
5. Making sure that subordinates find it easy to get help with their problems and complaints.

Checklist for Communications

For the purpose of analysis, it is useful to look first at:

- *The message.* What is this supposed to be? What language is it to be put in? What information does it contain?
- *Communicators.* Who are they? What are their roles? Where do they stand on the status scale? Is there a status gradient? Does either or both have a vested interest in communicating the message? Are their personalities likely to interfere with the communication process?
- *Media.* What form should be used? What are the mechanics of the information handling? What is the density of the communication system? What is the time pattern?
- *The environment.* What are the circumstances of the communication? Are they appropriate on this occasion? What about situational factors, especially who must know and who must not? Is there a protocol that is the social setting?
- *Effects.* How effective is the system? How capable is it of adaptation? What are its aims? Are they being achieved?

A suggested fault tree is shown in Figure 8-1.
A standard fishbone can be used, as shown in Figure 8-2.

ALCOHOL AND DRUG ABUSE

In a good program, an effective means of dealing with alcohol and drug abuse is a necessity. An effective program has the following characteristics:

- Employees with problems are not allowed to work and are perceived as being dealt with effectively by supervisors and the company program.
- The employees' program is visible and is credited with helping to eliminate alcohol and drug abuse on the job.
- Supervisors are trained in how to spot and how to deal with abusers—both what and what not to do.
- Educational programs are available to all employees.

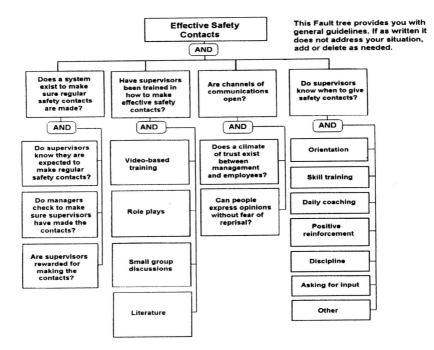

Figure 8-1. Fault tree for effective safety contacts.

There are no statistics on how many industrial injuries result from substance abuse, but it is undoubtedly a major factor. However, as big as the problem is, only recently have controls for it been developed in industry. The elements of a control program are discussed below.

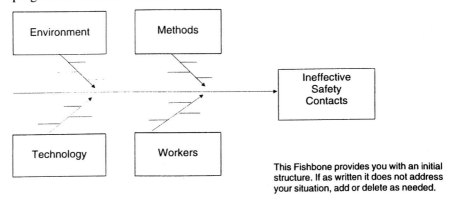

Figure 8-2. Fishbone diagram for ineffective safety contacts.

Preemployment Screening

Business groups of all types are evaluating different methods of screening drug users at the preemployment level. It stands to reason that drug users will avoid applying for jobs where drug screening is required. Therefore, if a company is not screening for drug users, drug-using potential employees will appear on the job site.

Employers have the right to know of any potential problem when hiring employees. Urine drug screening at the preemployment level is an excellent loss prevention method. Management has an obligation to provide a safe and healthy workplace for all employees and has the right to hire the most qualified applicants for the available job positions.

Identifying Problem Employees

Drug users sometimes are difficult to identify unless some types of urine or blood and plasma screening procedures are incorporated into a company's drug program. However, drug users do establish a pattern of unusual behavioral habits, which, if they are alert, company representatives can detect. Some of these symptoms include:

- An increase in quality control problems.
- Low production output.
- An increase in automobile liability premiums.
- Increased absenteeism.
- Worker's compensation rate increases.
- Signs of employee theft surfacing through inventory shortages.
- Morale beginning to deteriorate.

Supervisors are the first line of defense in combating employee drug abuse. Their observations come from working with the employees at the point of operation. Some of the signs they notice are:

- Some employees taking more breaks than others.
- Employees going to the restroom or their lockers in groups.
- A sudden change in individual personalities; short tempers where patience once existed.
- An increase in employee reprimands.
- A noticeable rise in employee complaints.
- Missing tools and equipment.
- Individual loss of work quality or production output.

- Employees sleeping on the job or showing up in areas where they do not belong.

An Employee Assistance Program (EAP) can be an integral part of the substance abuse program. Employees are referred to EAP counselors by supervisors for poor job performance, a positive urine screen test, or admitting a drug or alcohol problem.

Counselors will attempt to assist employees in solving their problems and to return a more productive worker to the supervisor. This task may be accomplished through group therapy, one-on-one sessions, or encounters with the employee's immediate family and support groups. Periodic monitoring of an employee's drug habit may be done in conjunction with the company's ongoing urine drug screening program.

The key to a successful drug abuse program is the development of a comprehensive policy which must be detailed and broad enough to address all situations likely to be encountered. It is important to define which employees will be covered by the policy. Employers should consolidate as many elements as possible into a single comprehensive policy, to avoid confusion when it is necessary to discipline employees. The policy also will provide excellent documentation in the defense of a legal challenge.

Employee Assistance Programs (EAPs)

Many of the problems that impact health and safety on the job originate outside the work environment. The changing family structure, single parents, and alcohol/drug concerns are some of the major societal problems that affect employers and employees, resulting in increasing absenteeism, accidents, increased costs of health care, or declining productivity. EAPs are emerging as one of the ways industry can reduce the cost of these problems.

The problems employees present to a comprehensive EAP include the following:

- Alcohol and/or drug problems.
- Marital and family relations.
- Emotional problems.
- Single parenting.
- Financial difficulties.
- Legal problems.
- Work-related problems.

An EAP should offer features such as:

- 24-hours/7-day-a-week service.
- Confidential services.

The Management System to Improve Worker Behavior

- Voluntary participation.
- Service to both employees and families.
- Being free of charge to the employees and their families.
- Coverage for any problem.
- Availability by appointment.

A suggested fault tree is shown in Figure 8-3.
A standard fishbone can be used, as shown in Figure 8-4.
A performance model can be used to assess individual performance problems (Figure 8-5).

AWARENESS PROGRAMS

Some organizations conduct campaigns to heighten employee awareness of safety and encourage the perception that the company's interest in safety extends to off-the-job as well as on-the-job activities. These programs seldom, if ever, affect behavior, but they are considered a part of the "tradition of safety." Be-

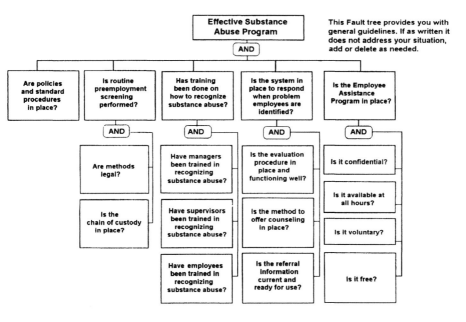

Figure 8-3. Fault tree for an effective substance abuse program.

122 THE AREAS TO ANALYZE

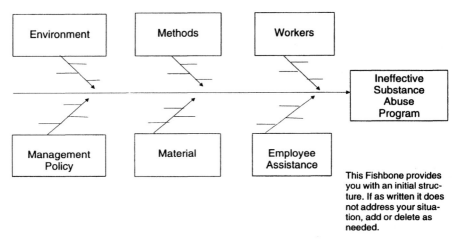

Figure 8-4. Fishbone diagram for an ineffective substance abuse program.

cause most companies engage in these activities, any organization should check to see the level of employee acceptance of these efforts.

The effects of behavioral influences can be illustrated by the chart in Figure 8-6.

An employee comes to the organization with certain influences from the past that dictate his or her behavior. These values are tested in the present situation, resulting in a "current motivation." This, coupled with his or her abilities, produces current behavior. We attempt to influence that behavior with a safety system, which consists of antecedents and consequences.

Antecedents (things that come before the behaviors) are rules, regulations, group norms, climate, and so on. Consequences are what the person experiences after his or her behavior. Although both are important, consequences are infinitely more important than antecedents. The rule (an antecedent) is only important insofar as it predicts what will happen following the act (consequences). Every employee knows which rules to follow and which to ignore after a short time on the job, learning from the reactions of the boss and others. Some of these are influences over which we have some control, while others are not. For those that we cannot control, it still is important to recognize that they are in fact influences.

Influences beyond Our Control

For practical purposes, all the past influences are outside of our control. So too are the following current influences: personality, attitudes, values, individual

The Management System to Improve Worker Behavior 123

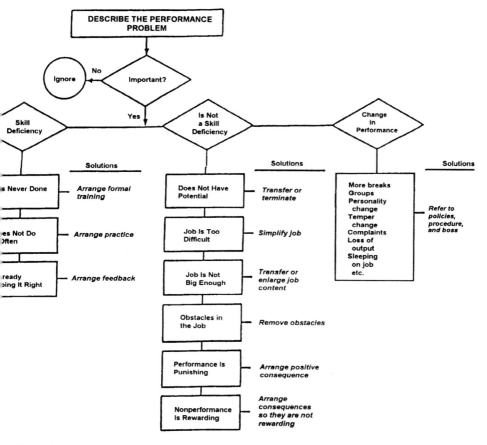

Figure 8-5. Performance model for evaluating performance problems.

differences, and the union. Building a management system that helps a supervisor better understand the worker's individual personality, attitudes, and individual value systems certainly helps to soften the effect of these influences on safety performance.

One study suggests that a system be provided to supervisors (after some training in concepts) to require them to look at each supervised individual and make a quick subjective evaluation of that person in terms of personality type, which theoretically is linked to the risk of the person's being in an accident. The supervisor then chooses a leadership style that seems most appropriate for the individual.

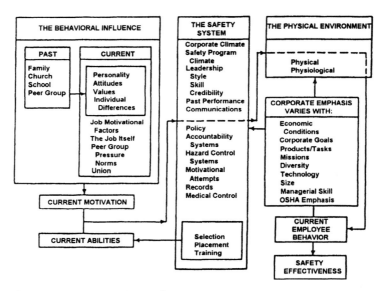

Figure 8-6. Effects of behavioral influences.

Influences within Our Control

Those influences within our control are considerably stronger influences than the ones beyond our control. They include the job motivational factors, the job itself, peer groups, the organizational climate, and supervisor styles. Improvement here, enhancing the motivational field of each employee, results from a systematic approach, probably administered by each employee's immediate manager. A start at this can be accomplished by providing supervisors with some simple methods to enable them to look at each of their employees.

Awareness Campaigns and Gimmicks

An integral part of traditional safety is gimmickry: contests, incentives, posters, hoopla. Does all this fit? Early on this writer became disenchanted with gimmickry for a very simple reason: it only worked sporadically. Thus, the comments here on gimmicks are somewhat biased. Gimmicks simply do not fit any place in safety theory; they make little sense in management theory and even less sense in behavioral theory.

- In management theory, performances are clearly defined to measure validity of performance and to make rewards consistent with their importance. Gimmicks define no performance; they usually measure with invalid measures

The Management System to Improve Worker Behavior 125

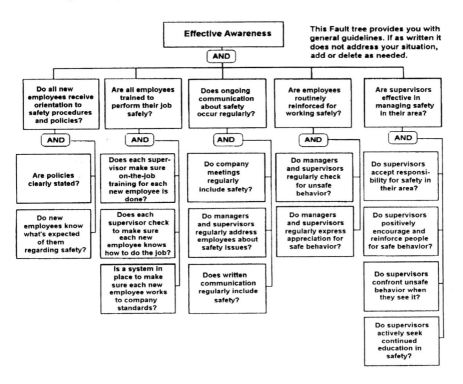

Figure 8-7. Fault tree for effective safety awareness.

(number of accidents) and reward with trivialities (plaques, tee shirts, jackets, steak dinners).

- In behavioral theory, a reward should be contingent upon performance to affect behavior, and the reinforcement must be contiguous (occurring right now) to the behavior. Gimmicks are neither contingent nor contiguous to any behavior.

In brief, gimmicks are irrelevant to behavior. This is not to say they should be eliminated. We must recognize that to eliminate gimmicks in most circumstances is to incur the wrath of almost everyone, from the old safety director to the lowest rated worker, for gimmicks are a satisfier. They do not affect behavior, but to eliminate them is to remove something expected. They maintain morale, but they do not shape behavior in most normal workers.

A suggested fault tree is shown in Figure 8-7.

126 THE AREAS TO ANALYZE

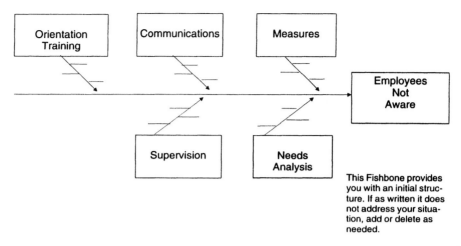

Figure 8-8. Fishbone diagram for use when employees are not aware of safety.

A fishbone is shown in Figure 8-8.

Chapter 9

The Management System to Improve Physical Conditions

Traditional safety emphasized physical conditions, and most regulatory approaches still concentrate on physical conditions. Are they important? Yes. How important? Heinrich said they "cause" 10 percent of our accidents. Current thinking does not ignore conditions; it only suggests that the presence of unsafe conditions tells us that something is wrong—the system is not working. In this chapter we speak of two of the 21 safety categories or areas (about 10%), but even these are presented only to look at the management system that allows them to exist.

INSPECTIONS

A good system includes regular inspections of all operations, with special attention to high-potential operations. Employees are aware that inspections are being made and may be involved in making them.

Inspections should include identifying safety hazards, health hazards, management system weaknesses, and the causes of human error. The category should also include system audits, intervention, surveys, or whatever is necessary to determine where loss can occur.

Inspection is one of the primary tools of the safety specialist. Today, however, one key question should be asked by every safety specialist engaged in inspection: "Why am I inspecting?" The answers to that question dictate how, when, and where to inspect. For instance, inspecting in order to unearth physical hazards means looking only at things. However, inspecting to pinpoint both physical hazards and unsafe acts also includes people. Unfortunately, most inspections today are of the former kind, rather than the latter.

If the primary intent is to detect hazards not seen before, the inspection differs from those cases where the primary interest is in checking on the inspections the department supervisor has made. If the intent is to detect hazards only, they can be corrected immediately by going directly to the maintenance department and reporting any deficiencies. If the intent is to audit the supervisor's inspection, the findings will be used to instruct and coach the supervisor so that future inspections will improve.

Many articles have been written on safety inspections and many have asked, "Why inspect?" Some typical answers have been:

- To check the results against the plan.
- To reawaken interest in safety.
- To reevaluate safety standards.
- To teach safety by example.
- To display the supervisor's sincerity about safety.
- To detect and reactivate unfinished business.
- To collect data for meetings.
- To note and act upon unsafe behavior trends.
- To reach firsthand agreement with the responsible parties.
- To improve safety standards.
- To check new facilities.
- To solicit the supervisor's help.
- To spot unsafe conditions.
- To measure the supervisor's performance in safety.

It is generally agreed that the condition and the people are the responsibility of the line supervisor. Thus, responsibility for the primary safety inspection must be assigned to the supervisor. "Primary safety inspection" means the inspection intended to locate hazards. Any inspections performed by staff specialists, then, should be done only for the purpose of auditing the supervisor's effectiveness. Hence the results of inspection become a direct measurement of safety performance or effectiveness. It is important to look behind the acts and behind the conditions when inspecting, and to ask, "Why are these present?" The answer to this question may lead back to the department supervisor, or it may lead to some other system weakness within the company; but the question should be explored fully and answered completely.

For instance, if an unsafe ladder is discovered, the inspector should immediately ask such questions as: "Why is this ladder here? Why was it not uncovered by our ladder inspection procedure? Why did the line supervisor

The Management System to Improve Physical Conditions   *129*

allow it to remain here?" Answers to questions such as these begin to get at the true causes of accidents.

The line supervisor often is provided with a checklist to use for the primary inspection, but this approach does not encourage an effort to trace the symptoms back to their true causes. This type of checklist should be reworked into a form that requries determination of some of the causes of the symptoms that have been unearthed.

Using the term "inspection" means to look at physical conditions only, overlooking the true causes of accidents as described in chapter 4, where human error was briefly discussed. Today it seems appropriate that inspections look at both physical conditions and worker behavior, to access the system weaknesses that cause them.

A suggested fault tree is shown in Figure 9-1.
A standard fishbone is shown in Figure 9-2.

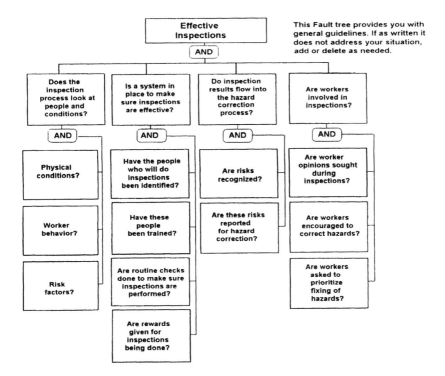

Figure 9-1. Fault tree for effective inspections.

130 THE AREAS TO ANALYZE

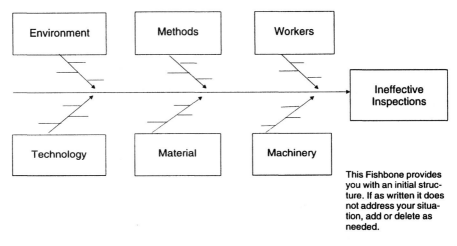

Figure 9-2. Fishbone diagram for ineffective inspections.

HAZARD CORRECTION

A good safety system has methods for dealing with reported hazards that are understood and supported by all levels of the organization.

Hazard correction includes the following:

1. Methods used to determine what hazards exist:
   (a) Who, how, and what.
2. Systems to prioritize those hazards through some systematic means that look at:
   (a) Frequency of expected loss.
   (b) Seriousness of loss from the hazard.
   (c) Cost of hazard removal.

This also includes how progress on hazard removal is regularly communicated to everyone in the organization.

Once hazards have been found, what happens next? To simply say "fix them" fails to deal with reality. In most environments there are ten times more hazards than can be fixed. It is easy to find them; it is expensive to fix them. In most companies a hazard-free environment would mean two things: no production and bankruptcy. Therefore, some practical solutions must be introduced to control hazards (a concept not always clear to enforcement personnel).

Some basic decisions must be made on what to fix and what not to fix. How much money should be spent, and what will the payoff be for the money spent?

The Management System to Improve Physical Conditions

These are management decisions much like any other management financial decisions.

There have been interesting economic treatises in this area. Notable is one by John Gleason and Darold Barnum, who looked at the management financial decision of whether or not to comply with OSHA standards. Here is an excerpt:[1]

> Analysis has indicated that expectancies of being cited for initial safety health violations, and the fine levels if cited, are so low under OSHA that they are of little value in preventing violations of the Act. Those employers who obey the law would do so regardless of the penalties. Employers at whom the sanctions are aimed—those who will correct violations only if it is economically profitable for them to do so—are not being affected. Thus the current sanctions antagonize employers who will obey the law only if it is economically profitable.

If we were to follow the above logic, it would make little sense to abate hazards (comply with OSHA standards). While economically this might be true, it is legally dangerous to follow this direction in the real world. At the other end of the decision scale is to totally abate all hazards (totally comply) regardless of cost effectiveness, which might delight enforcement officers but upset the financial wizards (who occupy crucial seats in most companies). Therefore, we must opt for some in-between course of action; we must prioritize our fixes.

Each organization will handle the problem differently. However, there are a few general principles that apply:

1. *Set priorities.* Most organizations cannot do everything at once; so decisions must be made about what comes first.
2. *Schedule and assign tasks.* Without planning and assigning responsibilities, not much will happen.
3. *Follow up* to ensure that things are being done.
4. *Document everything done,* if not for internal reasons (to protect yourself), certainly for OSHA.

Should the presses or the exit lights be rewired first? Should collapsible handrails for the loading dock or handrails around the roof be built first? These are

---

1. Reprinted from J. Gleason and D. Barnum, Effectiveness of OSHA penalties: myth or reality? Report from the Wisconsin Department of Industry, Labor, and Human Relations; in D. Petersen and J. Goodate, *Readings in Industrial Accident Prevention,* New York: McGraw-Hill, 1980.

the kinds of questions that someone will have to decide. For each item on the list there should be a start and a finish time.

Assess the capabilities of the maintenance people. Can they do this work as well as their regular tasks within the required time frame? Many organizations have found that they must set up a separate safety maintenance force. It can report either to staff safety or to maintenance as long as its efforts are directed solely to safety items.

Not until the organization is set can priorities be established and then a schedule set. One method of setting priorities employed a matrix:

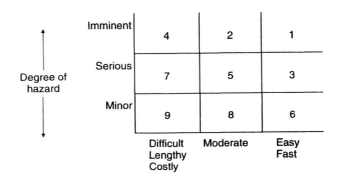

The matrix was used (1) by determining, as a compliance officer might, the degree of hazard (imminent, serious, or minor) and (2) by estimating the costliness of the change in terms of time, personnel, and money. The number in each box of the matrix—which indicates priority—was arbitrarily assigned at the outset. It could be shifted, depending on management's decision, money available, personnel, and so forth. As supervisors or inspectors reported to safety personnel any physical violations they could not correct themselves, the violations were given priorities, ranging from 1 to 9, and put on maintenance (task force) lists. That assured early completion dates for imminent and serious hazards (particularly those of moderate or small cost). Once priorities were established, the maintenance task force scheduled each job and set a start and a finish date. The schedule became the work assignment schedule for the task force, the report to management, and the needed documentation. Updated weekly, it showed the progress in compliance.

A number of systems are discussed in the safety literature, ranging from the simple matrix to sophisticated mathematical prioritizing.

A suggested fault tree is shown in Figure 9-3.

A standard fishbone can be used (Figure 9-4).

The Management System to Improve Physical Conditions 133

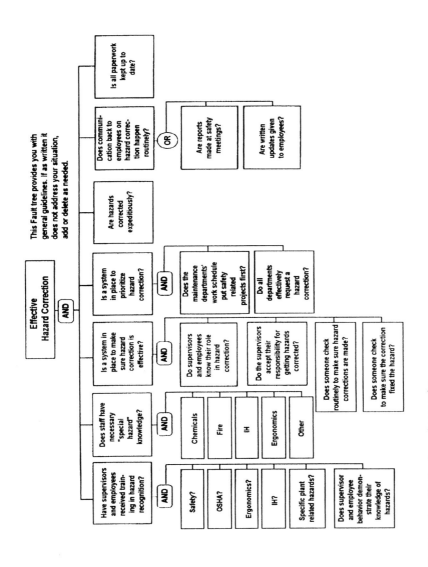

Figure 9-3. Fault tree for effective hazard correction.

134   THE AREAS TO ANALYZE

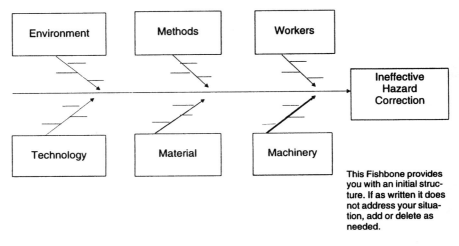

Figure 9-4. Fishbone diagram for ineffective hazard correction.

Chapter 10

Additional Areas to Analyze

In the last six chapters, 21 areas were discussed that need analysis through some process in order to determine effectiveness and to diagnose problems in the safety system. These 21 do not cover the entirety of what must be done in an organization to control loss, but they are important areas, perhaps core areas to safety excellence. They also are areas that perhaps can best be assessed by asking members of the work force for their opinion on how effective the system is.

This chapter briefly looks at other areas that usually must be covered, and that need to be assessed. In many cases the analysis might best be done by specialized people (safety professionals, medical professionals, human relations people, organizational development people, etc.). These areas include (but are not limited to) analysis done as follows:

1. In the management system for continuous improvement
    (a) Audits
    (b) Catastrophe control
    (c) Violence control
    (d) Vehicle accident control
2. In the management system to improve behavior
    (a) Shift work/overtime/overload
    (b) Behavior tracking
    (c) Safety in high performance organizations
3. In the management system to improve physical conditions
    (a) Purchasing and design
    (b) Ergonomics in reducing system-caused human errors
    (c) Ergonomics to lessen CTDs
4. In the management system to contain costs

Of course this list is limited, and could include much more. Some subjects are notable by their absence: OSHA compliance, industrial hygiene, occupational health, wellness programs, exercise programs, and so on. All are related to safety but are bodies of knowledge covered in depth by professionals in their field, or are really less related to safety than they originally appear, at least in the control approach.

In this chapter we examine the above list and some related topics.

IN THE MANAGEMENT SYSTEM FOR CONTINUOUS IMPROVEMENT

Audits—Is Yours Effective?

Since Audits were discussed earlier in Chapter 2, we mention them only briefly here, through the fault tree in Figure 10-1, where the key points are shown.

Catastrophe Control

Too often our safety systems, our regulations, and our recordkeeping focus on controlling all injuries with equal emphasis, regardless of their potential severity and financial loss to the organization. In the areas of regulation and required recordkeeping, we must count equally a bee sting and a fatality. At times a paper

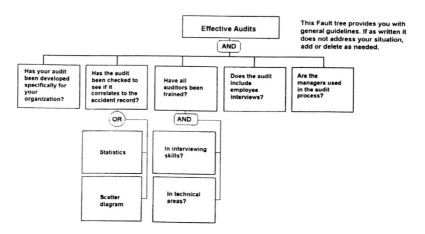

Figure 10-1. Fault tree for effective audits.

cut in the office is counted as a recordable injury, whereas an explosion that killed only members of the public and no employees is not recordable.

In our efforts it is probably wise to look specifically at how well we are controlling the areas that leave us vulnerable to a catastrophic event.

There is some research on the causes of catastrophes, particularly those caused by humans. One of the best investigations was the research by Edwin Zebroski of Aptech Engineering Services, who focused on a number of major human-caused catastrophes around the world to determine whether or not any common negative characteristics were present before the incidents occurred. As quoted in my book *Human Error Reduction and Safety Management,* some of his conclusions were as follows:[1]

> The most fundamental human factor is obviously the management—the capabilities, organization, and degree of involvement in proactive safety and reliability practices. Sometimes it takes a great catastrophe to bring the needed capabilities and involvement into play. Five recent examples—TMI, Chernobyl, Bhopal, the Challenger shuttle, and the Piper Alpha oil platform—are examined later in this paper based on an earlier study on the first four events and newer information on the fifth one. These events are important to study and to understand for several reasons. They help to clear the air on the risky attitude "it can't happen here." And they certainly help to make the point clear that "good practices" are not clichés. The basic lesson is that the absence (or the weakening) of just a few good practices can lead to catastrophe. It is not hard to make a list of the practices whose absence can make a catastrophe not only probable, but essentially certain. Then it is just a matter of how long it will take to happen.
>
> When we examine the common factors in large man-made catastrophes, we always find that many relevant and ultimately crucial factors were ignored. Nobody wanted to look at them or put them "on the table" for decision on the protective or remedial actions needed. One of the real benefits of structured decision analysis is to make some of the key factors in a decision *explicit* rather than implicit, and to get in view any "sacred cows" that affect and sometimes put blinders on how decisions are made or delayed.
>
> Eleven attributes that were found have had medium to large degrees of commonality in the basis for the TMI-2, Chernobyl, Challenger, and Bhopal events are:
> 1. *Diffuse responsibilities* with rigid communication channels and large organizational distances from decision-makers to the plant
> 2. *Mindset* that success is routine with neglect of severe risks that are present

---

[1]Reprinted from Lessons learned from man-made catastrophes by Edwin L. Zebroski, in *Risk Management,* pp. 51–65, by R. Knief et al., Hemisphere Publishing, Washington, DC, 1991. Reproduced with permission. All rights reserved.

3. *Rule compliance* and the belief that this is enough to assure safety
4. *Team player emphasis* with dissent not allowed even for evident risk
5. *Experience* from other facilities not processed systematically for application of lessons learned
6. *Lessons learned disregarded* and neglect of precautions widely adopted elsewhere
7. *Safety analysis and responses* subordinate to other performance goals in operating priorities
8. *Emergency procedures,* plans, training, and regular drills for severe events lacking
9. *Design and operating features* allowed to persist even though recognized elsewhere as hazardous
10. *Project and risk management techniques* available but not used
11. *Organization* with undefined responsibilities and authorities for recognizing and integrating safety matters

The matrix in Figure 10-2 lists the eleven attributes and identifies the catastrophe to which each was a major contributor. It may be observed that all of the systems basically share at least ten of the eleven attributes. Chernobyl was surely not excessively dependent on rule-following, since it had very few written rules and none for the test leading to the accident. Bhopal had written rules but was chronically out of compliance with many of them. The most solid exception to the eleven attributes was for Three Mile Island, where excessive devotion to production over safety was not a significant contributor to what happened.

MATRIX OF COMMON ATTRIBUTES
FOUR SEVERE ACCIDENTS

	BHOPAL	CHALLENGER	CHERNOBYL	TMI-2
Responsibilities	X	X	X	X
"Mindset"	X	X	X	X
Rule Compliance	(X)	X	O	X
Team Agreement	X	X	X	X
Prior Events	X	X	X	X
Narrow Experience	X	X	X	X
Output vs Safety	X	X	X	(X)
Severe Accidents	X	X	X	X
Known Hazards	X	X	X	X
Risk Techniques	X	X	X	X
Safety Integration	X	X	X	X

Figure 10-2. Catastrophe vs. attribute matrix. (Reprinted from Lessons learned from man-made catastrophes by Edwin L. Zebroski, in *Risk Management,* pp. 5–65, by R. Knief et al., Hemisphere Publishing, Washington DC, 1991. Reproduced with permission. All rights reserved.)

Additional Areas to Analyze 139

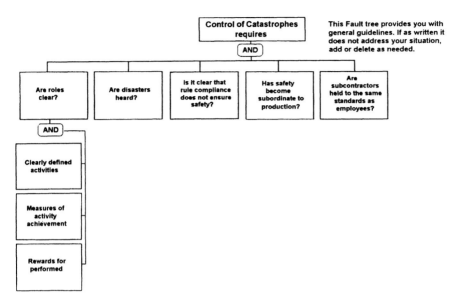

Figure 10-3. Fault tree for catastrophe control.

Some of Dr. Zebroski's conclusions are summarized in the fault tree shown in Figure 10-3.

Violence in the Workplace

The third leading cause of on-the-job fatalities in the United States is violence, an area seldom considered in traditional safety efforts. We tend to ignore it, perhaps regarding it as a security problem, not a safety problem; and yet many of the causes of our violence problems go back to the same underlying causes of our accident problems, our stress problems, our cost containment problems, and many others.

An article in *Professional Safety* magazine, by John W. Kennish, highlights some of the factors involved:[2]

> The Justice Dept. reports that nearly one million violent crimes—or almost one-sixth of all reported violent crimes in the country—occur in the work-

---

[2]Reprinted with permission from J. W. Kennish, Violence in the workplace, *Professional Safety,* June 1995.

place. It estimates that 8 percent of rapes, 7 percent of robberies and 16 percent of assaults occur in the workplace yearly. Therefore, workers may have a one-in-four chance of being the victim of some form of violence at work.

Most homicides result from arguments or disputes between two people who know each other well, such as co-workers. A seemingly simple verbal conflict may eventually escalate into a violent act. In many instances, those who witness adversarial verbal exchanges fail to see the danger of the situation; as a result, the problem remains and violence eventually ensues.

In many cases, companies fail to properly prepare for and respond to such conditions. Nearly 25 percent of the 311 companies surveyed by the American Management Assn. (AMA) said at least one of their workers had been attacked or killed on the job since 1990. But according to the same survey, only 24 percent offered some employees formal training on coping with workplace violence; just 10 percent offered such training to all employees. Fewer than 50 percent of the firms surveyed had procedures in place to deal with violent incidents, demonstrating an obvious lack of preparedness (Rigdon).

Increasingly, employees are venting their work-related anger and frustration through aggression and other violent behavior. A report from the National Institute for Occupational Safety and Health (NIOSH) stated that from 1980–89, nearly 12 percent (7,603 of 63,589) of workplace deaths were classified as homicides. For that period, homicide was the third-leading cause of occupational death in the U.S., behind motor vehicle crashes and machine-related incidents.

Violence specifically directed against employers and employees by other employees is now the fastest-growing category of homicide in the U.S. According to the American Society for Training and Development, rates of workplace homicide may have nearly tripled since 1989.

Data also reveal that workplace homicide is the leading cause of death for women in the workplace, accounting for 42 percent of all on-the-job fatalities.

NIOSH figures establish that 17 percent (1,072 of 6,308) of on-the-job fatalities in 1992 were attributed to homicide. During that year alone, approximately 111,000 incidents of workplace violence were reported. Not only did these incidents result in 1,072 deaths and thousands of injuries, but also an estimated $4.2 billion in related costs for employers.

Numerous civil actions and lower workforce morale and productivity resulted as well.

Interpersonal acts of violence—whether at home, on the street or in the workplace—have common variables and causes and are, to a degree, predictable. To better understand this problem, one must review violent actions in general, not as they relate to the workplace, but to people and their relationships. To understand and control workplace violence, motivation for such an attack must be identified, as must the type of person likely to commit such an act.

Violent incidents are likely the culmination of a series of "red flags"— early warning signs of the impending violent act. Recognizing these warning signs, and knowing how to defuse disputes early may significantly reduce the number of violent incidents—instead of just temporarily cooling off the combatants.

Common characteristics of violent acts include:

1) Violence generally begins as a verbal dispute and almost always involves persons who know each other.

2) Disputes often have relatively trivial causes, such as employee–supervisor arguments over business relationships.

3) The dispute pattern is usually not random and unique; rather, it builds, intensifies and continues.

4) Prior to violent incidents, eventual victims may behave in a provocative manner. More often than in other crimes, victims provoke the attack that ends in their injury or death. Frequently, they misjudge their own ability to arouse hostility in others, and are unaware that they are provoking a serious response to their behavior.

5) By far, men are more frequent offenders and victims. In *Breaking Point: The Workplace Violence Epidemic and What to Do About It,* the authors determined that the workplace killer was a solitary aggressor in 98 percent of 125 cases studied (Kinney and Johnson).

6) Availability of firearms contributes significantly to an eventual violent outcome. Because of their versatility and power, firearms are often weapons of choice in violent acts. Many attacks perpetrated with guns would simply not be committed with other weapons. Use of guns is a major determinant in whether a violent attack ends in homicide or aggravated assault.

7) Often the attacker has a police record and repeated prior adversarial police contact. Thus, a pattern of past violent behavior may be an accurate predictor of future violence. However, the previously noted AMA study found that only 48 percent of the companies surveyed screened job applicants for prior criminal records.

8) Violence is often preceded by feelings of resentment, jealousy, helplessness, anger and extreme stress, as well as by threats, intimidation, verbal challenges and insulting comments from the attacker.

9) Violence is often foreshadowed by threats; those made by problem employees should not be ignored or tolerated. One study found that in 50 percent of cases where force was used, one participant had first made a threat (Kennish 22). Before the 1991 massacre at the U.S. post office in

Royal Oak, MI, the gunman had warned co-workers that if he did not win his job back, he would make the office look like that in Edmund, OK, where a postal carrier killed 14 people in 1986.

10) Violence is frequently preceded by acts of pushing or other physical contact by the attacker and/or victim. A shouting match can escalate into a physical attack if a supervisor simply places a hand on the shoulder of an angry employee.

11) Substance abuse contributes greatly to the possibility of violence.

Violence can be caused by many variables and motivational factors. Common motivational factors include job losses resulting from layoffs, mergers and downsizing, and ineffective and insensitive management techniques. When workers lose their jobs or perceive or experience unfair management practices, not only may they lose their earning ability, but also their identity, friends and self-esteem. They may also develop feelings of extreme anger, helplessness and/or frustration.

To prevent and control workplace violence, management should:

- gain a better understanding of employee perspectives;
- identify potential problem conditions or relationships;
- effectively (and tactfully) resolve situations that warrant attention *before* serious incidents develop.

To accomplish these goals, management should:

1) Communicate senior management's decisive stance on workplace violence to employees. Educate employees about the issue, including the nature and availability of the company's training program (which should address dangerous conditions).

2) Recognize that supervisors must periodically assume the role of conflict managers. Therefore, training should teach supervisors better ways to perceive problem conditions and manage developing conflicts. *They should not be left to their own resources in responding to hostile employee situations.* Training should address danger signals, negative trends, human factor needs, dispute and conflict resolution, effective communication techniques and effective ways of dealing with problem employees.

3) Screen new employee prospects. Seek criminal history information. Do not hire problem employees.

4) Increase/improve communication between employees and management.

5) Conduct employee attitude surveys to uncover potential problems.

6) Develop a confidential process that allows employees to report threatening circumstances to management.

7) Provide a quick, fair, responsible way for management to react to

Additional Areas to Analyze   *143*

threatening conditions and effect positive change. Engage an outside behavioral risk professional if necessary.

8) Establish a confidential employee assistance program (EAP) to allow employees with personal problems to seek professional help.

9) Develop "people-sensitive" processes for potentially negative circumstances, such as termination and downsizing.

10) Identify the best avenues of action *before* problems arise.

11) Ensure that the company's various departments espouse the same position on workplace violence.

12) Establish an effective security program, including controlled access; personnel identification, policies and procedures; and equipment and facility design. Consider hiring professional security personnel.

13) Provide sensitivity and nonforce conflict-resolution training to security personnel.

14) Establish an effective criminal justice system liaison.

Kennish well summarizes the problem, the causes, and the solutions. (See Figure 10-4 for a fault tree for violence control.)

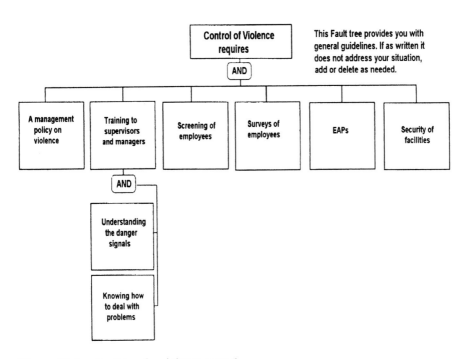

Figure 10-4. Fault tree for violence control.

Vehicle Accident Control

Vehicle accidents are the number one cause of on-the-job fatalities. As with violence, vehicle safety systems also are typically overlooked in traditional safety efforts. To ignore this problem is to ignore one of the major causes of injury and death. In short, the safety professional cannot afford to ignore the fleet of vehicles that his or her company runs.

This discussion is not intended for the safety professional who works with the true fleet operator, but rather for one whose responsibility extends to a small fleet of vehicles that is adjunct to the primary operations of the company. This might be the contractor's fleet, the manufacturer's pickup and delivery trucks, the fleet of bakery or dairy trucks, or the sales department's private passenger fleet. This section gives broad outlines of the accident controls found to be effective with this type of fleet operation.

One reason why the adjunct fleet is ignored is that safety staff are often so positioned in the organization that it is impossible to effectively reach and influence fleet management. Often safety staff report to industrial relations or to production management, whereas the fleet reports to an officer who is not closely connected with either industrial relations or production. The adjunct fleet may be a part of the sales department, or report to a traffic manager, to the head of material handling, or elsewhere. In terms of the principles expressed in Chapter 1, the safety professional who reports directly to the manager in charge of a particular activity probably will not be able to influence other areas, such as the fleet, easily.

Industrial safety specialists may not extend their influence to the fleet because they may not feel comfortable in fleet safety work; they may think that fleet safety is different from plant safety and that the approaches normally used to effect results will not work here. In fact, fleet safety is different from plant safety in two major ways, discussed below:

1. *Lack of supervision.* In the plant or on the construction job all employees are under constant supervision. On the road employees are not supervised; they are on their own. This is perhaps the biggest difference between plant safety and fleet safety. Since supervisory control is so much more limited in fleet safety, we must rely more heavily on some other controls—in this case, on deciding who will drive initially. Selection of personnel becomes critical in fleet safety.
2. *The environment.* In industrial safety we lean heavily on controlling the environment—on making it as safe as possible. In fleet safety we can control the condition of our own vehicle, but most of the rest of the driving environment is not under our control. The other driver, the condition of the other vehicle, and the road are uncontrollable. Much of this lack of environmental control can be balanced by the quality of the driver.

Additional Areas to Analyze

Consequently, the selection and the training of that driver are of paramount importance.

Although fleet safety is different in some aspects from industrial safety, the approach of the safety professional remains unchanged. The following are areas that might be looked at first; notice the similarity between these areas and those in industrial safety:

- Management's policy
- Driver selection
- Driver training
- Vehicle maintenance and design
- Records

Policy

A statement of management intent is as essential in fleet accident control as it is in plant safety. The policy in fleet accident control may be separate, or it may be an integral part of the total management safety policy, depending on the organization.

In either case the following points must be spelled out:

- Management considers safety on the road important.
- The corporate safety program will apply to the driver.
- Employee cooperation is expected.
- Specific responsibilities for safety have been assigned to the various levels of management.
- Accountability will be fixed.

Driver Selection

Selection is the single most important control that management has in fleet safety. Proper selection of drivers requires that management determine the abilities and skills of applicants for the driving job. To do this, it attempts to obtain information on the driver's experience and performance on previous driving jobs, job knowledge (technical know-how), and attitude toward safety. Management should also consider the driver's job performance during the probationary period, which is often overlooked.

The specific tools of selection are:

- The application form.
- The interview.
- The reference check.

- The license check.
- The physical examination.
- Written tests.
- Road and yard tests.

In the screening process, management's first job is to determine how applicants have lived and worked in the past. This is the best single indicator of how they will live and work in the future. Of the above list of selection tools, those concerning past performance are the most important indicators.

One of the best tools is the application form, as it covers the essential facts:

- Driving experience—local, long haul.
- Job performance record. Does the applicant stay with a job?
- Responsibility, maturity, and stability.
- Past safety performance.

The personal interview provides a face-to-face encounter with the applicant. Here the interviewer can appraise the person's knowledge, attitude, character, and maturity. The purpose of the interview is to gather facts. In the interview, these matters ought to be discussed:

- Driving experience.
- Knowledge and education.
- Knowledge of the vehicle.
- Experience with the vehicle, maintenance.
- Record—arrests, violations.

After the application and the interview, some reference checks should be made. Checking prior work references establishes the validity of the information obtained and can be done simply and inexpensively over the telephone. The following questions should be answered:

- Was the applicant employed by the company as stated?
- What type of work was he or she engaged in?
- What was the applicant's absentee record?
- What was his or her wage record?
- What were the applicant's reasons for leaving?
- Would the company rehire this person?

In addition to making the reference check, management should ensure that the applicant has a valid driver's license. Also, a routine check with the applicable state agency for past violations is well worthwhile.

Every applicant should be given a physical examination to determine whether

Additional Areas to Analyze   *147*

he or she meets the physical requirements for driving. The driving job, perhaps more than most other jobs in industry, requires this physical check. It seems inconceivable that any manager would put an expensive vehicle, and the possibility of a million-dollar lawsuit, into the hands of a person who cannot see or is subject to blackouts. And yet this happens daily. It happens even in companies with sophisticated safety programs, where those programs do not seem to apply to the salesperson in the company car.

The application, interview, and physical examination are the central part of any selection system. Management would do well to ensure that the potential employee comes through these steps well before adding some of the "frills."

Various kinds of tests in addition to the above are often valuable. These tests can take many forms and generally include those questions that the company thinks its drivers should be able to answer. Standardized tests are also available. For example, the yard test is used to determine a driver's skill in handling the equipment without going into traffic. Some exercises often included are the parallel park and the alley dock. This test does not predict whether a person will be a good driver—only whether he or she can maneuver the equipment. The road test (in traffic) should be made over a predetermined route which approximates the kind of driving that the applicant will be required to do if hired.

Driver Training

After a driver is finally put on the payroll, he or she must be oriented and then trained. Most of the principles discussed in the section on training as a tool of motivation apply to driver training also. The training given should, as much as possible, aim at defined needs and should accomplish stated training objectives. Objectives are essential to quality training. Basically, the content of a training program is what management wants the workers to know.

Orientation might include:

- Company policies and practices.
- State, county, and local traffic laws.
- Defensive driving.
- Customer or public relations.
- Concepts of safety.

The ongoing training of drivers might include the following:

- Vehicle operation.
- Vehicle condition, including how to check the vehicle daily and what to report.
- Use of company forms.

- Emergency procedures, including roadside warning devices, use of fire extinguishers, proper use of accident report forms, witness cards, conduct at the scene of an accident.
- Demonstrations, in the use of the brake detonator, skill-exercising equipment, and other training devices.
- Psychophysical testing devices. The validity of these lies in their training value, not in their use as measurement devices. They can be used to train drivers on reaction time, night vision, and stopping distance.
- Films and visual aids.

Vehicle Maintenance

Fleets usually evolve their own systems of preventive maintenance, which are devised to fit their special needs. Maintenance of leased fleets varies according to the contract between lessee and lessor. A fixed-cost or full-maintenance arrangement usually includes maintenance, repair, and insurance. A financed lease is one in which the leasing company supplies the vehicles, and the lessor conducts the maintenance. Regardless of the setup, a preventive maintenance system must be used. It is the role of the safety staff specialist to ensure that whatever the arrangement for maintenance, the program will be effective.

Generally some kind of maintenance record system is necessary to ensure effectiveness. Records should be designed to:

- Show when maintenance is needed.
- Provide a schedule of what needs to be done.
- Record what has been done.
- Give costs.

Usually, the fleet maintenance record system uses the following forms:

- *Driver's condition report*—a checklist of things to be checked daily by the driver and an order for necessary corrections (to be sent to the shop).
- *Maintenance scheduling form*—a means of ensuring that the shop schedules all vehicles periodically for routine service.
- *Service report*—a record showing detailed inspection by mechanics when the vehicle is in for routine service and also repairs needed and when made.
- *Vehicle history card or folder*—a complete history of each vehicle.

Probably the best sources for comprehensive recordkeeping systems and forms for preventive maintenance are those of the truck manufacturers and oil companies.

Additional Areas to Analyze   *149*

Records

Two other types of records necessary in fleet operations are driver records and accident records.

Driver Records

Every fleet needs to keep some records on each driver. The minimum is a file containing information on:

1. Hiring
   (a) Application form completed for screening
   (b) Character references
   (c) Results of work reference checks
   (d) Record of violations and accidents
   (e) Results of interviews
   (f) Results of tests given
   (g) Physical examination report
   (h) Information on previous driver training
2. Job performance
   (a) Supervisor's reports
   (b) Any commendations
   (c) Status in company's safe-driving award program
   (d) Performance in maintenance functions
   (e) Record of training received
   (f) Vehicles driven or assigned
   (g) Losses of cargo, money, etc.
   (h) Any road observation reports
   (i) Any violations of warning notices
   (j) Accidents
   (k) Property damage reports
   (l) Complaints
   (m) Appraisals

Accident Records

Accident records start with the driver's submitting an accident report. In the case of serious accidents, investigations are usually made by the company and the insurance carrier. Management may wish to collect, analyze, and summarize these accident reports to determine trends. Often the information and/or reports are evaluated for type of accident, immediate causes, and driver chargeability. Causes

150 THE AREAS TO ANALYZE

are generally classified in terms of driver, roadway, and vehicle. Of primary importance is proper analysis of driver-related causes. Usually, analyses list improper driver actions as:

- Failure to yield right-of-way
- Following too close
- Failure to signal intentions
- Excessive speed
- Failure to obey traffic signals or signs
- Improper passing
- Improper turn
- Improper backing
- Wrong traffic lane

Some of these ideas are summarized in the fault tree in Figure 10-5.

IN THE MANAGEMENT SYSTEM TO IMPROVE BEHAVIOR

Shift Work/Overtime/Overload

A consideration often overlooked in management decisions is the influence of shift work and overtime on the accident record. Much has been written and researched in this area, with most investigators suggesting that there is a strong

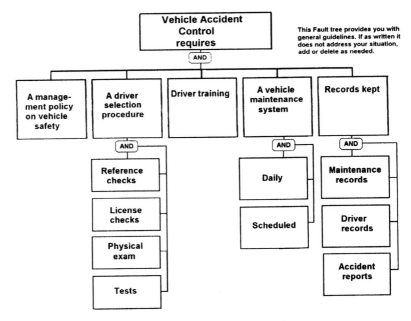

Figure 10-5. Fault tree for vehicle accident control.

Additional Areas to Analyze 151

negative effect on the behavior of affected employees, and thus on the accident record. Some of this research was recorded in my book on *Human Error Reduction and Safety Management,* which is used as a resource here.

Overload comes from the improper matching of capacity with load, which takes place in a particular state. When this occurs, a person may be overloaded or underloaded, either condition being dangerous. The element of state is important, as people can be overloaded severely for a long period, and yet with the right attitude, or the right motivation, can survive and thrive. Without that attitude or motivation only a small overload can be disastrous (the straw that breaks the camel's back).

Overload can be physical, physiological, or psychological. Because physical and physiological overload are dealt with in a number of other places (OSHA standards, industrial hygiene texts and standards), we primarily will look here at psychological overload and its outcomes.

Overload is more prevalent today than ever before in organizations. As Adrienne Burke has written:[3]

> Overtime hours are at a historic high now, according to the Bureau of Labor Statistics. In September, workers in manufacturing industries averaged 5.1 overtime hours weekly. A Bureau of National Affairs survey in July found that 90 percent of responding employers occasionally required overtime of workers. Of continuous operations (running 24 hours a day, 7 days a week), nearly half admitted to frequently mandating extra hours.
>
> Stories pointing out overtime dangers abound. Interviews for this report uncovered these: A maintenance worker is so fatigued he fumbles around looking for a wrench and finds it in his other hand. After four days of 18-hour shifts, a utility line worker electrocutes himself. A nurse accidentally sticks herself with an HIV-infected needle in the twelfth hour of a 12-hour shift.

A lot of research has been done on shift work and sleep. For example, the National Commission on Sleep Disorders research shows that insomnia, the inability to sleep or sleep well, affects an estimated 40 percent to 80 percent of shift workers. On average, a shift worker sleeps two to four hours less a night than a day worker in the same age group. According to the National Commission on Sleep Disorders, people are generally sleepiest between 2 A.M. and 5 A.M. There also is a dip in alertness between 2 P.M. and 3 P.M.

With the rightsizing efforts of many (most) organizations, the whole area of shift work, overtime, and overload is perceived by many to present a serious safety problem. The fault tree in Figure 10-6 offers us some direction in this area.

---

[3] Reprinted with permission from Adrienne Burke, The push to produce, *Industrial Safety and Hygiene News,* December 1994.

152   THE AREAS TO ANALYZE

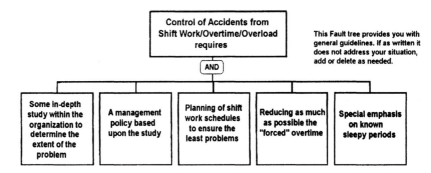

Figure 10-6.   Fault tree for accident control with shift work/overtime/overload.

Behavior Tracking

Systems to improve worker behavior are of crucial importance in today's safety concepts. We believe that concentrating on behavior offers our best bet for an excellent safety record. Whether we believe in the old Heinrich axiom claiming that 88 percent of all accidents are caused by unsafe behaviors, or the modern theory asserting that in 100 percent of the incidents there has been a human error, we must place major emphasis on the improvement of safe behaviors.

In the process of behavior improvement it becomes imperative that we measure our progress by behavior tracking, and not by accident records or system audits. Appendix B details the process of behavior sampling as a measurement of performance.

Figure 10-7 shows a fault tree for the assessment of the behavior tracing process.

Safety in High Performance Organizations

During the last ten or fifteen years, management has undergone some amazing transformations. The changes in management thinking have been fast and furious in recent years; as each new book is published, management changes its direction and philosophy on how to manage.

The transition in management thinking started a number of years ago when managers began to question the dictates of classical management, or scientific management, as described in the early years by Frederick W. Taylor.

Following the classical school of management thinking came the human relations school of thought. To a degree the human relations approach, popular in management until the 1970s and still taught today, was an outgrowth of misinterpretation of the research of the 1920s, '30s, '40s, and '50s. Many interpreted the

Additional Areas to Analyze   153

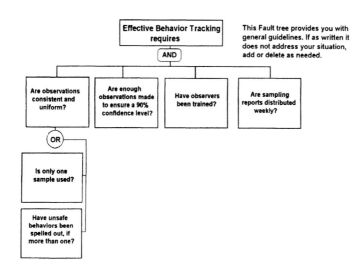

Figure 10-7. Fault tree for effective behavior tracking.

research to say that "happy" workers are productive workers; thus the role of management was to make workers happy. Actually the research did not say it quite that way, as the subject is considerably more complicated than that.

The underlying assumptions of the above two schools of management thought were similar but slightly different. Classical management is based on the assumption that "Everybody is alike—we can get the behaviors we want through manipulation." Thus we used wage incentive plans, piece rates, and other schemes to get more productivity. Human relations management has the underlying assumption that "Everybody is alike—we can get the behaviors we want by making them all happy." Neither assumption has any basis in fact.

In the 1970s the contingency school of management thinking emerged, and for the first time, management's underlying assumptions about people were in tune with psychological reality. The assumption behind the contingency theory of management is that "Everybody is different," so that our management style, and how we deal with a worker, must be contingent upon the situation, the workers, and their needs. How a manager manages must be appropriate to the situation.

The situational leadership approach of Hersey and Blanchard became very popular in the late 1970s and early 1980s when one of the original authors, Blanchard, restated the basic thesis in different terms in the book *The One Minute Manager,* which became a best seller.

The cultural assessment approach to management became popular in the 1980s with the publishing of *In Search of Excellence,* one of the most popular management books in years. The authors quite simply looked at the U.S. companies that consistently seem to be the best performers, the most effective in terms of bottom-line performance; and from this in-depth look at the best-run companies, they distilled some keys to effective management.

Up to this point in the management philosophy evolution, we, in safety, could remain fairly comfortable, for we could still try to feed our classical management-style safety programs into our organizations. Then came the 1990s, and new and different trends in management became popular. The Deming philosophy and the total quality management concepts began to threaten our comfort levels, for now we were talking employee empowerment—no rule setting by management—using statistical process control tools that could literally make many of our old safety tools obsolete.

In addition, rightsizing and reengineering in some companies had the effect of putting more and more work on fewer and fewer supervisors, managers, and employees.

In this process, one of the emerging trends has been to go to high performance organizations to decide that there are levels of our traditional hierarchy that we can do without. In some organizations this has meant the flattening of the organizations—reducing the number of middle managers and the number of layers of the hierarchy. This approach makes great sense in an age of computerization and engineering. In flattening the structure, we have become more efficient, more productive.

In other organizations, a decision has been made that we no longer need first line supervisors and should go to team management. This decision has great historical underpinnings—we have known since the old Hawthorne studies of the 1920s that the most productive work units are those where there is no formal supervision. It might be remembered, however, that in these early studies, team leadership was only an experiment—and, as a result, people responded very positively (exhibiting the Hawthorne effect, as we call it today).

As we go to team control from supervisor control, how do we ensure an accident/illness-free operation? The team approach requires us to reassess and reunderstand how safety can be accomplished in any organization.

The Total Involvement Concept

First, the high performance team concept needs some definition. It is different from what we have been talking about in recent years when discussing worker participation, involvement, or empowerment. The team concept suggests that decisions will be made and problems will be solved by work teams, not by individuals, as in empowerment.

Empowerment, participation, and involvement can all happen in a hierarchical structure when management simply asks the workers to help in managing the

Additional Areas to Analyze   *155*

organization, or one function of it, such as safety. In a traditional but worker-empowered organization, little is changed with respect to safety. We still have clear accountability within the management structure for proactive daily performance of those things that prevent incidents although workers may do many of them. But accountability remains clear—it remains in line management from CEO to first line supervisor.

The Team Concept

In the team concept, however, things have changed. First, there is no line supervisor to be held accountable through the traditional management tool of measurement coupled with rewards. With regard to safety, there are other things to be considered.

Traditional Safety Principles in High Performance Settings

Let us consider a few of our key safety principles in a team environment:

- The foreman is the key man (Heinrich, 1931): Today, there is no foreman.
- There are three Es of safety (adapted from Heinrich, 1931)—engineering, education, and enforcement: Two of these were traditionally the foreman's responsibility.
- Safety should be managed (Petersen, 1970).
- The key is accountability (Petersen, 1970): How do we measure and reward a team?
- Management starts with a clear definition of roles: What is the role of the team?
- We must decide who will do the traditional supervisory tasks of inspecting, investigating, motivating, training, one-on-ones, behavior observations, etc.

Our Attempts to Date

As we jumped into high performance approaches, we seem to have opted for one of several approaches, based on these assumptions:

- We assumed that the teams would automatically take care of safety along with everything else.
- In many cases we also assumed that by merely removing a person (the supervisor), teams were ready for self- (team) management.
- We assumed that we in management should treat the team as we would a supervisor, saying that "You are responsible for your own safety; now do something about it."

- We assumed that with employees now in charge of safety, management was "off the hook," and no longer had a role—they had abdicated.
- We assumed that all people were ready to "take over." However, some were not; some are happy just to do their thing.
- We assumed that all employees wanted a piece of the action. However, some did not. We created a three-tiered organization: (1) management, (2) involved workers, and (3) uninvolved workers, creating in some places another friction point.
- We sometimes assumed that if it did not work, we could always go back and take back the authority; but when we did so, it resulted in considerable chaos (a mild description).

High Performance Successes

Occasionally the team concept worked brilliantly, resulting in unheard-of success. Having identified some of the pitfalls, we now will discuss how to achieve success with team approaches.

Many organizations have been highly successful in utilizing the team approach. I have seen small units that have made the transformation with remarkable results. In talking with team members in successful companies in a variety of industries (food processing, oil drilling, railroads, steel companies), I have never seen such excitement from hourly workers as from those experiencing team participation. And, without fail, the resultant safety records have been superb. I also have seen the opposite—organizations where going to team approaches has produced problems and discontent, reduced morale, increased friction, and occasionally led to utter chaos.

What are the differences? Probably a number of things: readiness for participation, the planning that went into the change, the amount of involvement the workers had in the change decision and process, and so on.

In those organizations that have achieved the desired results, here are some steps that management has seemed to follow—things that any organization might well consider:

1. Define within management exactly where you are willing to go by defining what you mean by participation. Is it to get better input so management can make better decisions? Is it to share decision making? Is it to turn decision making over to the teams? And which decisions will management retain?
2. As well as possible, check the maturity level both of management and the worker. Find out if the organization is "ready." This can be done though interviews or perception surveys. Also, look at the degree that self-management has been allowed in the past before going to teams.

Additional Areas to Analyze

3. Check the amount of confidence and trust that exists between management and the workers. Culture surveys can help in the assessment.
4. Clearly define the ground rules, the guidelines defining what is allowable in decision making and what is not (regulatory compliance, core corporate values, etc.).
5. Allow great flexibility within those guidelines. For instance, "You must have a system on your team of behavior observations, but how you do it is up to you."
6. Require each location/team to have an annual (or periodic) plan in safety. Have an annual (or periodic) check of each unit to see whether or not it is carrying out the plan.
7. Hold each location/team accountable for what it has agreed to do.
8. Determine the core elements that should be dealt with in each plan, but allow great flexibility in what each location/team can do to achieve results within those core elements.
9. Require intermediate self-checks in shorter intervals, making frequent adjustments as needed.

See Figure 10-8 for a fault tree on high performance safety requirements.

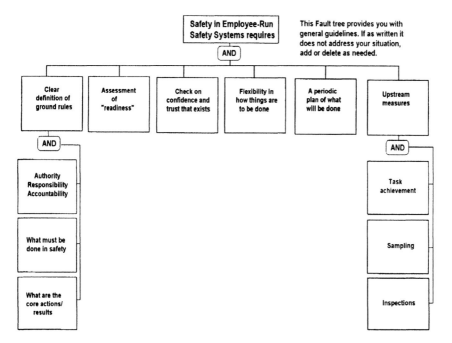

Figure 10-8. Fault tree for safety in employee-run safety systems.

In the Management System to Improve Physical Conditions

Purchasing and Design

Although we believe today that safety primarily involves behavior, we also think that many incidents are due to system-caused human error. Often we design equipment or purchase equipment and materials that lead to injuries and loss.

Part of the solution to this is an ergonomics problem—to identify where we have in the past designed in safety problems. Part of the solution lies in purchasing and design.

Purchasing

In purchasing, the staff specialist can do a good job of analyzing present controls and spotting weakness. Some of the considerations are:

- What specifications are there for purchasing?
- Are the purchasing specifications set by engineering? Are they based on product tests and known standards?
- How are incoming parts or materials tested?
- Are proper records kept of these tests?
- Are materials or incoming parts identified so that they can be traced to the supplier?
- What procedures ensure that design changes are communicated to purchasing?
- What is the warranty setup with suppliers?

Design

Loss control is limited in design and engineering because the controls are primarily a function of the qualifications and performance of the design staff. Although the safety professional is not competent to make any judgment on such matters, certain questions can be posed:

- Have written job descriptions and standards of performance been established for engineering and design personnel? Are they regularly reviewed?
- Does the head of the engineering department know whether each employee is practicing the things necessary to produce a good design?
- Is all work checked by a superior before release to manufacturing?
- Do all product failures receive a full analysis so that design and manufacturing can make any necessary changes?

Additional Areas to Analyze

- Do design and engineering personnel receive training? If so, what is the purpose of such training?
- Are rigid specifications written for each part? Are drawings and specifications tied together so that purchasing, manufacturing, and quality control follow the desires of the original design?
- Are all changes in materials and manufacturing approved by design?

See Figure 10-9 for a fault tree.

Ergonomics in Reducing System-Caused Human Errors

"Ergonomics" is a common term today in industry, and even beyond industry. It appears to be a politically correct term from the amount of publicity it receives. However, ergonomics seems to mean only one thing: how we can change work stations so people will move their bodies less and thus reduce Cumulative Trauma Disorders (assumed to be due to repetitive motion).

Ergonomics is regarded as a new field; yet when I studied industrial engineering in the early 1950s, I studied exactly the same material that is popular today, only

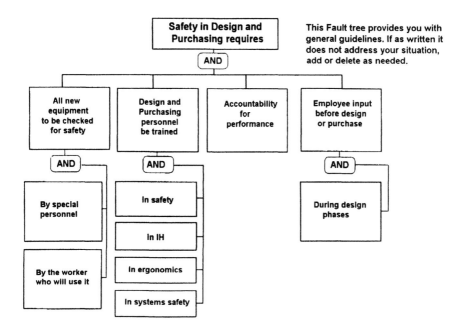

Figure 10-9. Fault tree for safety in design and purchasing.

from a productivity standpoint. We called it time and motion study, and learned how we could design work stations to meet the anatomy and stereotypes of humans, so that they could reduce their motion and thus be more productive. As popularized today, ergonomics is extremely limited—it looks primarily (only) at designing to reduce motion, to reduce the probability of (perhaps) repetitive-motion–caused injuries. In this discussion we will split ergonomics into two parts: first, design intended to increase efficiency and reduce human-caused errors, and, second, the efforts we use to reduce repetitive motion.

There are many books and manuals on ergonomic-systems analyses, and we will not attempt to reproduce the material here—it clearly is way beyond the purview of this book. As a teaser, however, we provide the chart shown in Figure 10-10, which suggests some aspects of such analysis.

Ergonomics to Lessen CTDs

Much has been written in the last decade about Cumulative Trauma Disorders (CTDs) related to repetitive bodily motion. OSHA prepared comprehensive guidelines a number of years ago that never actually became a standard or a guideline, yet major fines were promulgated under the General Duty Clause.

Later different guidelines were floated, which included the following:

Scope Applies to all employers where there is daily exposure to:
—Performing same motions (2, 3, or 4 hours)
—Performing awkward posture (2, 3, or 4 hours)
—Using vibrating tools (2, 3, or 4 hours)
—Using hand exertions (2, 3, or 4 hours)
—If any employee has a problem

Application Multi-employer workplaces
(temps and subs)
Host and service companies share the responsibility

Identification of Problem Jobs
Score of more than 5 on the risk factor checklist (easy to get 5)
Must tell employees the score, symptoms, how to report

Control—Must bring each job down to 5
—Must have weights on boxes (over 25 pounds)
—If cannot get to 5:
—Fix within 60 days or
—Job improvement process (in writing)
1. Perform job analysis
2. Implement control measures
3. Fix within 60 days

Additional Areas to Analyze

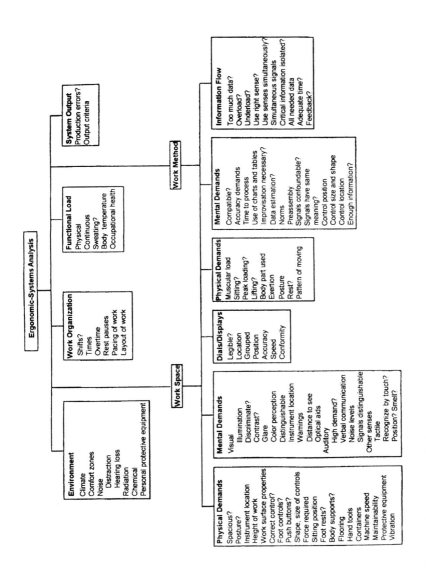

Figure 10-10. Ergonomic-systems analysis.

—New Jobs
  1. Risk factor analysis within 60 days
  2. Controlled within 60 days
—Training
  1. For each employee who analyzes
  2. Awareness training to each employee
  3. In all languages of employees
  4. With proof of training effectiveness
  5. During work hours
  6. Initially and annually
—Medical Management
  1. Contact of person with healthcare provider
  2. Each affected employee to have a health management plan
—Recordkeeping
  1. Records on problem jobs
  2. Records on the job improvement process
  3. Records on training
  4. Records on medical management
  5. Records made available to OSHA

Another approach would be to schedule a look at the jobs in the organization (hopefully by line supervisors or employees), using the chart shown in Figure 10-11, and to assess to what extent there are potential problems on these jobs.

In all of the suggested approaches, to a large degree the human factors involved in CTDs have been ignored. Some are described by John Kamp:[4]

> Employees engaged in physical work have always suffered aches and pains. However, until 10 years ago, these employees could take only two courses of action: 1) keep working despite the pain or 2) find another occupation. Given the "self-diagnosed" nature of disability in most cumulative trauma cases, however, these employees now have a third choice: filing a WC claim.
>
> Psychological factors, job attitudes and personal traits play a key role in whether this becomes the option of choice. The role of job attitudes is straightforward. Consider Susan and Dorothy, two employees who perform the same physical work in neighboring factories. In addition to equal physical exposure, these employees also suffer an identical, moderate level of pain.
>
> Satisfied Susan loves work—she gets along well with her supervisor, her best friends are at work, and she feels her employer cares about her welfare.

---

[4] Reprinted with permission from John Kamp, Worker psychology, safety management's next frontier, *Professional Safety,* May 1994.

Hand & Wrist	Minor Concern	Major Concern
Grasp • pinch grip • static hold	• brief pinch grip	• prolonged pinch grip • forceful grip
Wrist Posture • flexion/extension • ulnar/radial deviation	• flex/extend 20–45°	• flex/extend > 45° • radial/ulnar deviation
Frequency • hand or wrist manipulations	• 5–10 per minute	• > 10 per minute
Mechanical Stress • localized pressure to palm or fingers • scraping/bumping • strike with hand • single finger trigger	• brief exposure	• prolonged exposure
Vibration • high frequency vibration	• brief exposure	• prolonged exposure

Neck	Minor Concern	Major Concern
Neck Posture • bend/twist > 20°	• < 50% time	• > 50% time

Back	Minor Concern	Major Concern
Lifting	• 75%–90% of AL	• > 90% of AL non-NIOSH
Torso Posture • torso bending • torso twisting	• bend 20–45° • twist > 20°	• bend > 45° • bend > 20° & twist
Static Hold/Carry • > 5 seconds	• 5–10 lbs.	• > 10 lbs. • 5–10 lbs. with flexed shoulder
Static Load • not able to change sit/stand posture over work day	• good posture	• poor posture
Push/Pull • whole body action vibration	• any exposure	• poor condition exposure

Arm & Shoulder	Minor Concern	Major Concern
Arm Work • exertions > 5 lbs.	• brief exertion	• little rest between exertions • large exertion
Static Load • prolonged holding	• supported	• unsupported • large exertion
Elbow Posture • fully flexed • fully extended • rotated forearm	• briefly	• repeatedly
Shoulder Posture • flexed • extended • abducted	• flex/abduct 45–90°	• flex/abduct > 90° • any extension
Mechanical Stress • sharp edges • hard surfaces	• brief exposure	• repeatedly • high force

Legs	Minor Concern	Major Concern
Foot actuation • foot pedals	• any	• excessive force; extreme posture; high frequency of duration
Leg Posture • knee • ankle	• slight squat	• deep squat • kneeling • 1-legged posture • walk/stand on uneven surface
Mechanical Stress • localized pressure • kicking	• brief exposure	• high force • prolonged exposure

Figure 10-11. Jobs ergonomic analysis chart.

Conversely, Disenchanted Dorothy hates work—her boss always berates her, her work unit is full of animosity and tension, and she feels her employer is merely using her. Clearly, continuing to work despite the pain is a more attractive option to Susan, while staying home and collecting workers' compensation has greater appeal to Dorothy.

Certain personal traits increase the likelihood that an employee will file a cumulative trauma claim, even when another employee with a similar pain level and job attitude would not. Hypochondria, the tendency to constantly adopt a "sick person" role, is one such trait. This behavior is reinforced by attention from doctors, family, friends, etc. Being "out" on workers' compensation provides an opportunity to be lavished with attention.

Another trait that contributes to cumulative trauma claims is a poor work ethic. Certain individuals view work as a temporary annoyance (rather than one of life's responsibilities) to be tolerated only until a more desirable source of funding becomes available. When these individuals experience work-related pain, workers' compensation offers an attractive way to "make a living." (In the extreme, the trait of dishonesty leads some to file fraudulent claims in the absence of pain or injury.)

Research supports the role of psychological factors such as stress and job attitudes in WC injuries. For example, a prospective study of 3,000 Boeing workers found that, apart from a history of back problems, the strongest predictors of reported back injuries during a four-year follow-up period were not physical factors, but rather worker perceptions and psychological traits. "Subjects who stated that they 'hardly ever' enjoyed job tasks were 2.5 times more likely to report a back injury ($p = .0001$) than subjects who 'almost always' enjoyed their job tasks."

A NIOSH study examined factors related to upper extremity musculoskeletal disorders and symptoms among U.S. West Communications employees who used video display terminals (VDTs). Psychosocial variables associated with disorders and symptoms included increasing work pressure, surges in workload, lack of job diversity, limited decision-making opportunities, uncertainty about job future, fear of being replaced by computers, and poor supervisor and co-worker support. According to the researchers, "This study adds to evidence that the psychosocial work environment is related to the occurrence of work-related upper extremity musculoskeletal disorders."

St. Paul Fire and Marine Insurance Co. has conducted studies of relationships between stress, job attitudes and WC claims. These studies have utilized Human Factors Inventory (HFI), an employee survey that measures various "human risk factors" in an organization's workforce. Factors include work stressors (i.e., excessive workload), personal life stressors (family problems) and lifestyle risks (substance abuse) that increase organizational costs through accidents, injuries, health problems, and lowered productivity, quality, and customer service.

Additional Areas to Analyze

Table 10-1. HFI scores and WC data

Location	WC Freq.	WC Claim Coasts	WL	SP	WR	PP	PLS	JS
Low claims	1.9	$921	59	43	41	52	44	51
High claims	15.5	$192,770	77	58	59	66	55	65

Note: HFI scores of 50 are average; higher scores are more unfavorable.
WL = Workload; SP = Supervision, WR = Working Relations;
PP = Personnel Practices; PLS = Personal Life Stressors;
JS = Job Satisfaction

Table 10-1 shows HFI results and WC data for low-claims and high-claims locations of a nationwide copy machine distributor. As shown, the 15 low-claims locations, reported minimal WC frequency rates and costs in the year prior to surveying employees. The three high-claims locations reported an average standard frequency rate eight times higher than the low-claims locations with average costs nearing $200,000 per 100 employees.

High-claims locations scored significantly worse ($p < .05$) on many HFI scales, indicating greater workload pressures, poorer supervision and working relations, more dissatisfaction with personnel practices, greater personal life stress and greater overall job dissatisfaction. Within this company, therefore, high levels of work, personal stress, and unfavorable job attitudes were associated with higher WC claim rates and costs.

Table 10-2 shows HFI results and WC data for four California locations of a nationwide cable television company. California is the nation's leader in compensating purely mental job injuries (so-called stress claims). Therefore, by separating these locations' WC claim rates according to stress and non-stress claims, one can uncover a clear picture of trends. Rates reflect claims spanning the 15 months preceding the survey.

Traditionally, functions such as hiring, employee surveys, stress management, personnel policies, and EAPs have been the domain of the human

Table 10-2. HFI scores and WC data

Location	WC Claim Freq., past 15 mo.			Avg. HFI[a]
	Stress	Non-stress	Total	
A	2.5	9.4	11.9	48
B	1.6	17.9	19.5	51
C	8.2	28.3	36.5	61
D	12.6	28.8	41.4	62

[a]Average score across the 6 HFI Work Stressors scales.
Note: HFI scores of 50 are average; higher scores are more unfavorable.

resources professional, not the safety professional. However, the role of the human element in WC claims has become too important for the safety professional to ignore.

Figure 10-12 is a fault tree for control of CTDs.

IN THE MANAGEMENT SYSTEM TO CONTAIN COSTS

Although this book concentrates on accident prevention, we cannot ignore the other aspect of the injury problem, after the injury has occurred. In many cases (rightly or wrongly) our safety system is judged by the bottom line (dollars), and there are a number of things we can do to influence dollar outgo even though accidents have occurred. This cost may or may not be under the purview of the safety department; but, either way, it needs to be controlled.

A number of steps can be taken ahead of time that will help immeasurably in containing costs. Here are a few:

1. Have one person oversee the worker's compensation program.
2. Evaluate the facility for "light duty" productive work areas and jobs. This identification becomes crucial later.
3. Carefully select a company physician. You will find doctors who:
    (a) Charge a lot or not very much.
    (b) Are injured-employee-centered or company-centered.
    (c) Are knowledgable or not about worker's compensation.
    (d) Are knowledgable or not about your plant.
    (e) Know or do not know about the real world.

Here are some tips on appropriate selection:

- Consult with other employers in the vicinity about their recommendations on which physicians are most helpful and most experienced with occupational injuries and illnesses.
- Communicate with your worker's compensation insurance company, which can provide assistance and recommendations for selection of qualified and capable doctors.
- Select physicians, surgeons, and specialists who do worker's compensation work willingly and will abide by a fee schedule if there is one established.

The selection process requires:

- Meeting with and interviewing the doctor.

Additional Areas to Analyze

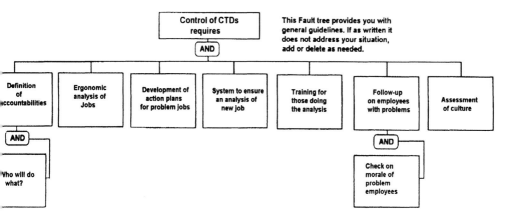

Figure 10-12. Fault tree for control of CTDs.

- Making sure the doctor tours and understands the facility, process, and culture of the company.
- Carefully selecting an insurance company/insurance claim adjustment service.

If you are insured, look carefully at the claim department policies and abilities before deciding on the carrier. This is crucial in cost containment efforts. If your claim representative (whether you are insured or self-insured) uses an open checkbook (never fights), your are heading for trouble. If you use an adjustment service (are self-insured) and you do not know what is going on or have no say in what is going on, you are in trouble also. In short, you must be in control.

After the event has occurred, here are some actions that can help:

- Ensure getting immediate qualified care. If the supervisor is required to take the employee for treatment, doing so shows care and maintains some control.
- Ensure that the accident investigation looks for third party recovery possibilities.
- Maintain periodic contact with the claim person on the case. Claim people respond to those customers who are interested. If an employee is off for three weeks, and no target return-to-work date has been provided, instruct the

		Employee's Name	Date of Injury

EMPLOYEE DISABILITY LOG

(Follow the steps below to obtain needed information when an employee has sustained an On-the-Job Injury)

W E E K #1	• Report accident/injury immediately.	
	• Contact employee.	
	... Date of doctor's appointment: _____	
	... Doctor's name: _____	
	... Date released to return to work: _____ ... Date of actual return to work: _____	
	• Contact employee's doctor.	
	... Date examined: _____	
	... When can employee return to regular job/alternate productive work?	
	... Describe injury: _____	
	... Contact insurance claims representative. Exchange information.	
#2	• Contact employee. Date contacted: _____	
	• Alternate productive work available?	
	• Return to work target date: _____	
#3	• Contact employee. Date contacted: _____	
	• Contact doctor: _____	
	... Return to work date: _____	
#4	• Contact employee. Date contacted: _____	
	• Contact insurance claims representative: _____	
	... If no target return-to-work date, insurance claims rep./plant management should schedule an independent exam.	
#5	• Contact employee. Date contacted: _____	
	... Advise of and instruct employee to keep independent exam.	
	... Date employee last saw doctor: _____	
#6	• Contact employee. Date contacted: _____	
#7	• Has employee been evaluated by our doctor for return to work?	
	• Results of exam: _____	
	• Contact employee: _____	
	... Confirm return to work date by phone and letter to employee and insurer.	
#8	• If employee hasn't returned to work, contact employee. Date contacted: _____	
	• Contact insurance claims representative regarding status of claim.	
	• Return to regular job tasks, or: _____	
Second Eight Weeks Continue with this Weekly Follow-Up For:	• Return to alternate productive work or: _____	
	• Examination with our doctor for return to regular or alternate productive work or:	
	• At such time that it is obvious that employee absolutely cannot return to work at your facility, work closely with the insurance claims representative and consumer loss control department to evaluate other alternatives:	
	e.g. Job placement outside; retraining for job placement.	

> Employee injury follow-up is a location management responsibility.

Figure 10-13. Employee disability log.

Additional Areas to Analyze 169

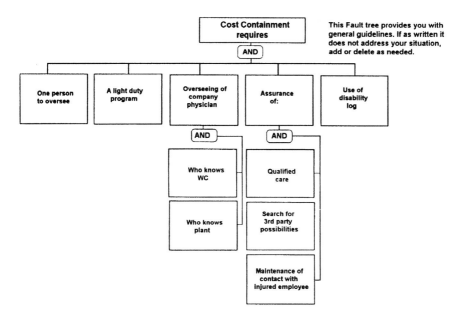

Figure 10-14. Fault tree for cost containment.

insurance company to set up, *immediately,* a second-opinion examination with a doctor of your choice. Purpose of exam: to obtain a return-to-work target date.
- Use a light duty program.
- Use a disability log (see Figure 10-13).

Figure 10-14 provides a fault tree for analysis of this area.

PART *III* | *The Change Process*

Chapter *11*

Interpreting the Data

In Chapter 2 we discussed some ways (mostly traditional) to analyze the effectiveness of a safety system, including such things as statistical assessment, audits, checklists, and so on. In Chapter 3 we introduced other approaches to analysis, including asking or observing the hourly employees, using perception surveys, interviewing, and behavior sampling. Then, in Part II, we looked at the areas to analyze.

No matter which approaches you use, you will obtain data that must be sorted out and analyzed to see what it means.

As we looked at the areas to analyze in Part II, we also suggested some ways to interpret the data in the 21 areas suggested: fault tree analyses and group solving approaches using fishbones and other SPC tools.

At times, however, the data collected contain much more information about what is right and wrong with systems than we can get by looking at an individual category alone. At times poor or good scores in several categories could well mean that something might be very good, or seriously wrong, in the entire management (or safety) system.

This chapter suggests some additional techniques for analyzing the data. Most of what follows will discuss the interpretation of perception surveys although it also pertains directly to the interpretation of interview-generated data and data from sampling, audits, checklists, and perhaps even statistics.

After a survey (of any type) you usually end up with reams of data—computer runs, charts of all kinds, and probably more information than you can possible digest. This chapter has to do with how you can handle all of this data, and use it to achieve:

- A "feel."
- A baseline.

- A diagnosis.
- An assessment of management's perception of reality.

Different companies have obtained different insights from perception survey results. Let me give you some examples, all from the real world:

1. One company, a large railroad employing about 20,000 employees, used the perception survey as a part of an experiment, but found it to be an excellent diagnostic tool. In analyzing and interpreting the results of the survey, these problems were immediately apparent:

- Recognition was a serious problem. Employees needed to be recognized for doing a good job, and they were not: 53 percent did not think they received the recognition due them. Management did not ever know the problem existed; in fact, 77 percent of management thought they were doing a good job of recognition.
- After recognition (or lack of it), the next three most serious problems, according to the employees, were: supervisory training, quality of supervision, and the company's inspection procedures (a function of supervisory responsibilities).

The survey results were cystal-clear: supervisors in the organization were not doing their job. They either did not know how to perform safety tasks, or they were choosing not to do them in the light of everything else they had to do each day.

2. In a second organization, a large steel company, the chart showed a different configuration. Again, a look at the overall results of the survey, listed in descending order of hourly perception, showed these weaknesses in the safety system:

- The lowest-rated area for hourly employees was the substance abuse program of the company. Management had studiously avoided confronting this issue for many logical and legal reasons. The employees, however, were saying loud and clear, "Please deal with this substance abuse problem; we no longer want to work with people who pose a daily hazard to us."
- The second-lowest-rated area was employee recognition (which emerges over and over again in almost all companies). Workers were pleading for positive feedback.
- Here again, supervisory performance rated low in all of the areas (quality of supervision, training of supervisors, and inspections), suggesting a lack of supervisory accountability (it made no difference whether supervisors did any safety work).

3. In a third company (another large railroad), the lowest-rated safety program area was "Safety Rules." Because this response was different from the norm of all

Interpreting the Data   *175*

companies surveyed to date, it needed further analysis. We needed to understand why "Safety Rules" were perceived as a serious problem by the work force. The category of safety rules in the perception survey comes from the response to three questions. Each was looked at to understand the low ranking of this category.

- Q: Are safety rules good for you? Here 79 percent responded favorably—no problem.
- Q: Are there too many safety rules? Here 80 percent responded unfavorably—no problem.
- Q: Does following safety rules interfere with production? Some 82 percent said yes, reflecting a serious problem.

This was a question of management's priorities. People were getting pressured for production, and the safety rules stood in their way! The problem was not the rules, but rather the perception at the employee level of management's priorities.

In this company, the second lowest employee rating was (again) for recognition, indicating the need for a system that forces positive recognition. The third, fourth, and fifth lowest employee ratings were for quality of supervision, supervisory training, and inspections (a function of supervision), again indicating the need for more supervisory training and/or supervisory accountability—a system to make it happen.

Perhaps the above examples give an indication of what can be gleaned from perception survey results. Again, the results can be used in a number of ways:

1. To get a "feel" for what people are thinking, primarily at the hourly level. The survey tells you where your problems are. The overall survey "score" tells you where your safety system is right now.
2. To provide you with a baseline of where you are right now. The percentage of favorable responses at the *employee* level in each of the 21 categories gives you a snapshot at a particular point in time of the effectiveness of your safety system in the 21 areas. Note that your baseline must be the hourly response only! Do not include any managerial responses in your baseline. This is extremely important for these reasons:
   (a) Research conclusively shows that management's opinion expresses what they "think" is happening or what they "wish" is happening, whereas employee response shows what *is* happening.
   (b) Managerial response usually tracks about 15 percent more positive than employee response (or more), trapping you into a false sense of "OKness."

In short, in using a perception survey you have to believe (because it is true) that hourly perception is reality.

3. To diagnose. With the belief that hourly perception is reality, looking at hourly responses tells you which areas of your safety system are working and which are not. Those with the lowest hourly positive response are not working. With this diagnosis, you know exactly what to fix in your safety system—corporately, by division, by region, by function, and so on. Goals can be set and activities defined at each level to fix what is not working.
4. To assess management's perception of reality. With the knowledge that hourly perception is reality, the survey results then also allow you (or force you) to assess management's (your) perceptions of what is real. Because those in management normally think that things are about 15 percent better than they are, if you find in a specific one of the 21 areas (or overall) that you are more than 15 percent apart from hourly perception, that suggests a serious problem—you and your management are not facing reality.

STEPS TO SURVEY RESULTS INTERPRETATION

> Step 1—Look at the lowest employee response categories.

Start with an overall look (total company). Start with a chart that shows all the categories for the whole organization surveyed, listed or charted in *descending* order of *hourly employee* response:[1]

This will show you your overall systemic problems. For instance, on the chart (Figure 11-1) the worst category is #1, Employee Recognition; only 47 percent of the employees think they are receiving the recognition they deserve for doing a good job.

In this chart three other categories are notably weak: #2, Supervisor Training, 50 percent favorable; #3, Quality of Supervision, 50 percent; and #4, Inspections, 52 percent. All three are areas that should be targeted for improvement.

As a general rule, anything below 60 percent positive needs to be targeted for improvement. In this case #5 (Management Credibility), #6 (Safety Regulations), and #7 (Support for Safety) should be assessed.

A discussion of some of the steps in interpretation follows.

---

1. The data in the following examples was compiled before the addition of the "Stress" category.

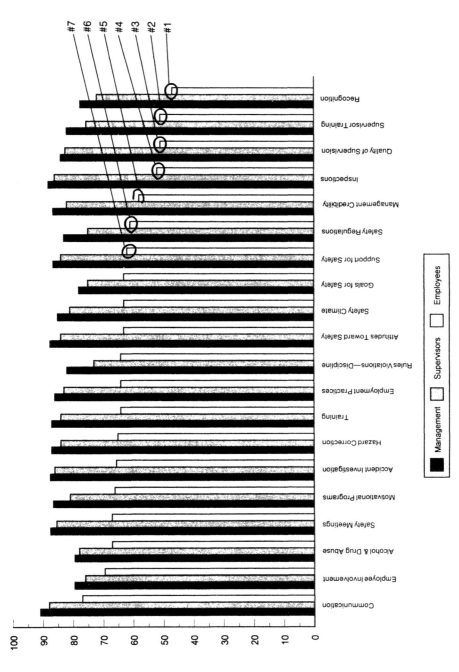

Figure 11-1. Survey response chart.

> Step 2—For further insight in a category, look at individual question response scores.

For further understanding of the response to any individual category, you may wish to look further. As the category response score is composed of the scores from the individual questions that compose that category, sometimes it is necessary to look at the responses to the individual questions. This is particularly helpful if the score does not make sense to you.

In the above chart the category Safety Regulations seemed low. In another company in the same line of work, it was the lowest category of all. This needed clarification: Was it a case of too many rules and regulations? Too few? Unfair enforcement?

Breaking down this category into individual question responses helps.

> Step 3—Look for clusters.

Identify the clusters of categories that help you to diagnose. In Figure 11-2 there are two clusters of categories with low scores at the employee level, which should indicate a serious problem to management.

Cluster #1

Cluster #1 (Figure 11-2) consists of the three categories Supervisor Training, Quality of Supervision, and Inspection. All three of these speak to the perception of the performance of the first-level supervisor, who clearly is seen as not doing a good job. Hourly employees are saying this supervisor is not competent, or is not trained sufficiently well to know how to do the job, or simply does not exert the time and effort to do what is necessary in safety. Normally this will mean a lack of supervisory training (supervisors do not know what they are supposed to do), or more commonly there is no accountability system requiring them to do the tasks they should. A quick analysis of the company's training efforts, and how supervisors are rewarded and measured for their safety performance, will give you answers to this very serious problem.

In some cases this cluster's scoring low might also show a serious flaw in the company's structure; it could mean that supervisors perceive themselves more as members of the work force than representations of management. This often happens in certain industries, railroads, construction, and steel, where tradition and collective bargaining have dictated that some first-level supervisors

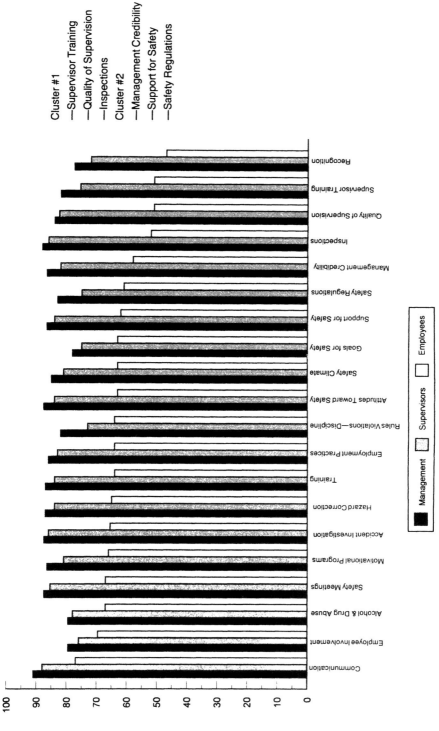

Figure 11-2. Clusters in the categories.

are not management—are clearly labor-union members, hourly rated, and so forth. At times the position is temporary; one day a person is a supervisor, the next a laborer.

Cluster #2

Cluster #2 (Figure 11-2) consists of at least two other categories: Management Credibility and Support for Safety. At times a third category (Safety Regulations) may be added. Both the Credibility and Support low scores indicate another serious problem—the workers just do not believe that management is interested in safety; they think that safety is a low priority in the organization. As indicated before, a low score in Safety Regulations could mean that management has made it clear that safety rules interfere with or slow production, and that production is number one.

Other Clusters

Other clusters to look for in your analyses are low scores in:

- Safety Regulations and Rules Violations–Discipline, which suggest an oppressive management, a them versus us climate.
- Employment Practices and Training, which suggest a lack of orientation and training—employees may not know what is unsafe. A low score in Practices with a high score in Training might suggest that they do not care to use what they know.
- Attitudes and Employment Practices, which suggest that they do not care.
- Safety Regulations and Employment Practices, which suggest that you have way too many rules; nobody is following them.
- Inspections and Hazard Correction, which suggest that physical conditions are deteriorating, and the problem is seen as management's fault. A high score in Inspections and a low one in Hazard Correction suggest a maintenance department problem—they cannot get around to fixing everything that is wrong.
- Employee Involvement and Communications, which suggest mushroom management.
- Climate and Regulations, which suggest either an oppressive management or one that does not care.

Look for groups of categories that tell a story about the reality of your organization.

Interpreting the Data

> **Step 4—Look for inconsistencies.**

Look for inconsistencies. Often things will not seem to fit together, for instance:

- A high in Communications with a low in Safety Meetings—maybe you no longer need meetings.
- A high in Involvement with a low in Safety Meetings—again maybe you no longer need meetings.

A stand-alone low for Alcohol and Drug Abuse means that employees are asking for management to do something!

> **Step 5—Compare yourself to norms.**

Compare your overall chart to the industry norms; over time these norms will become increasingly meaningful as the data base grows. See if there are categories in your organization that deviate significantly. For instance, as mentioned earlier, one steel company scored very low in substance abuse compared to others. One railroad scored very low on safety regulations, requiring a closer look.

> **Step 6—Compare to your baseline.**

Compare yourself to your baseline. The first survey establishes your baseline; later surveys will show your progress. The succeeding surveys will show you overall trends and whether or not your specific goals and programs in targeted categories are working.

> **Step 7—Compare to your other indicators.**

Compare your results to the other indicators of safety system success—audits, statistics, behavior sampling, and so on. In this comparison, keep in mind that this perception survey is your most valid indicator.

Compare the survey (what are the strengths and weaknesses of your system) against what your audits say. This tests the validity of your audit. When audit scores are high and perception survey scores low, your audit is not measuring what it should be, or what is important. The audit needs a thorough reexamination and testing.

Compare the survey results against your accident record. If the survey results are worsening over time while your accident record is improving, check your accident record input. It is possible that incidents are going unreported, or that there has been an increased effort in back-to-work programs. If you are surveying multiple locations, run scatter diagrams or make correlated studies comparing perception results by location and the reported statistics.

For instance, in one organization the comparison between perception survey results and accident statistics looked like this:

Rank Order Survey Results	Rank Order Accident Frequency
1. Location A	1. Location F
2. Location B	2. Location A
3. Location C	3. Location B
4. Location D	4. Location C
5. Location E	5. Location D
6. Location F	6. Location E

Further analysis showed that the manager of Location F had been under considerable pressure from corporate management to improve safety performance. This was done by reporting less than half of the injuries, which went undetected until the perception survey results came in.

Step 8—Look at the difference in perception between levels.

Next examine each of the categories (starting with the worst employee perception scores) and determine the difference between employee scores and management scores (see Figure 11-3). For instance, in the figure these differences stick out as serious problems:

Inspections—Management Perception	87% Favorable
Employee Perception	50% Favorable
Difference	37%
Quality of Supervision—Management	82%
Employee	49%
Difference	33%

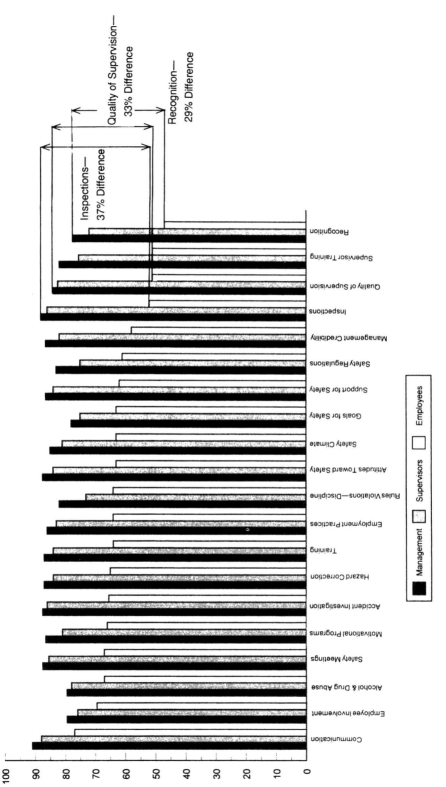

Figure 11-3. Determining the difference between employee and management scores.

Recognition—Management 76%
  Employee 47%
  Difference 29%

Not only are these three problem areas, but management is not aware of the problems.

When you use a perception survey, you must interpret it in the belief that the *employee perception is reality.* When members of management perceive things differently, they simply are not seeing reality. The percentage difference indicates the degree to which the problem exists. As a rule of thumb, anything over 15 percentage points should suggest a serious problem.

ANALYSIS BY UNIT

Now you are ready to make your analysis of individual subunits in the manner in which your survey was set up—by location (region, division, etc.) or by function (manufacturing, shipping, transportation, etc.).

The analysis is done in the way described above, following the eight steps:

1. Look at the lowest employee response categories.
2. For further insight in a category, look at individual question response scores.
3. Look for clusters.
4. Look for inconsistencies.
5. Compare yourself to norms.
6. Compare to your baseline.
7. Compare to your other indicators.
8. Look at the difference in perception between levels.

Chapter *12*

Where Do You Want to Be?

This question, "Where do you want to be?," suggests that the people in each organization must know where they are trying to go. They must have a vision of what their safety system should look like; and they must have an idea of what basic principles will undergird their system, or what criteria their safety system must meet.

This determination of a vision is not easy; it requires some deep soul-searching and thought about what you believe about safety, particularly regarding two main questions: What do you believe causes accidents? What do you believe must be done in your organization to control accidents?

As indicated in Chapters 1 and 2, safety and safety management have slowly but certainly been changing over the last century, with the changes becoming more rapid recently. A part of that change has been in what we believe about how accidents are caused and about how they can be controlled.

Our first real causation model was Heinrich's "domino" theory. From the very beginning we in safety have recognized that certain conditions are involved in accident causation. Thus as soon as it became economically feasible to try to control accidents, we immediately started working on physical conditions. Then in 1931 Heinrich informed us of what now is a painfully obvious and simple truth—that people, not things, cause accidents.

Heinrich stated it this way:

> The occurrence of an injury invariably results from a completed sequence of factors, the last one of these being the injury itself. The accident which caused the injury is in turn invariably caused or permitted directly by the unsafe act of a person and/or a mechanical or physical hazard.

He likened this sequence to a series of five dominoes standing on edge. These dominoes are labeled:

1. Ancestry or social environment.
2. Fault of a person.
3. Unsafe act or condition.
4. Accident.
5. Injury.

Most safety people have preached this theory many times. Many of us have actually used dominoes to demonstrate it; as the first one tips, it knocks down the other four unless at some point a domino has been removed to stop the sequence. Obviously, it is easiest and most effective to remove the center domino—the one labeled "unsafe act or condition." This theory is easy to understand; it is also a practical approach to loss control. Simply stated, it says, "If you are to prevent loss, remove the unsafe act or the unsafe condition."

The domino theory was intended to be a very practical system for removing the things that cause accidents. Perhaps, however, our interpretation of the domino theory has been too narrow. For instance, using the investigation procedures of today, when we identify an act and/or condition that "caused" an accident, how many other causes are we leaving unmentioned? When we remove the unsafe condition that we identified in our inspection, have we really dealt with *the* cause of a potential accident?

Today we know that behind every accident there lie many contributing factors, causes, and subcauses. The theory of multiple causation states that these factors combine in random fashion, causing accidents. If this is true, our investigation of accidents ought to identify as many as these factors as possible—certainly more than one act and/or condition. When we look at an act and a condition, we are looking only at symptoms, not at causes. Too often our narrow interpretation of the domino theory has led us only to accident symptoms. If we deal only at the symptomatic level, we end up removing symptoms and allowing root causes to remain and thus lead to another accident or some other type of operational error.

Over the years we have tended to become confused concerning our definition of accident causes. We have considered unsafe acts and conditions to be causes of accidents, and the things that allowed the acts or produced the conditions we have thought of as personal factors or subcauses. Many times we have looked for the "proximate cause" of an accident, as some laws and codes use this terminology.

One fact, then, seems clear: When we look only deep enough to find the act or the condition, we deal only at the symptomatic level. This act or condition may be the "proximate cause," but invariably it is not the "root cause." To effect permanent improvement, we must deal with root causes of accidents.

THE MANAGEMENT SYSTEM AND ACCIDENTS

Root causes often relate to the management system. They may be due to management's policies and procedures, supervision and its effectiveness, or training. In our example of the defective ladder, some root causes could be a lack of inspection procedures, a lack of management policy, poor definition of responsibilities (supervisors did not know they were responsible for removing the defective ladder), and a lack of supervisory or employee training.

Root causes are those whose correction would effect permanent results. They are those weaknesses that not only affect the single accident being investigated but also might affect many other future accidents and operational problems.

The root causes of accidents (weaknesses in the management system) are also the causes of other operational problems. Though this fact is not immediately obvious, the more we consider it, the more obvious it becomes. Consider, for instance, how often our safety problems stem from lack of training—and how often our quality problems also stem from this same lack of training. Or consider how poor selection of employees creates safety problems and other management problems. The fundamental root causes of accidents are also fundamental root causes of many other management and operational problems.

The Key Person

Another fundamental tenet of safety has stated that the supervisor is the key in accident prevention. This seems axiomatic in our thinking. The supervisor is the person between management and the workers who translates management's policy into action. The supervisor has eyeball contact with the workers.

Is this the key person? In a way, yes. However, although the supervisor is the key to safety, management has a firm hold on the keychain. It is only when management takes the key in hand and does something with it that the key becomes useful. Safety professionals have sometimes used the key-person principle to focus their efforts on frontline supervision, forgetting that the supervisor will do what the boss wants, not what the safety specialist preaches.

The above very brief description of accident causation concepts and accident control concepts will perhaps lay a small base for the reader to pursue what he or she perceives to be of value in these two areas.

THE FUNCTION OF SAFETY

It is only in recent years that most safety professionals have been able to define their role in the safety work that is being accomplished. What they do has changed and will continue to change as concepts and principles continue to evolve. If permanent results can be effected by dealing with root causes, safety professionals must learn to work well below the symptomatic level.

If accidents are caused by management system weaknesses, safety professionals must learn to locate and define these weaknesses. They must evolve methods for doing so. This may or may not lead them to do the things they did in the past. Inspection may remain one of their tools—or it may not. Investigation may be one of their tools—or it may not. Certainly safety professionals must use new tools and modernize old tools, for their direction is different today; and their duties must also be different.

In the safety profession, we started with certain principles that were well explained in Heinrich's early works. We have built a profession around them, and we have succeeded in progressing tremendously with them. And yet in recent years we find that we have come almost to a standstill. Some believe that this is so because the principles on which our profession is built no longer offer us a solid foundation. Others believe that they remain solid, but that some additions may be needed. Anyone in safety today ought to at least look at that foundation–and to question it. Perhaps the principles discussed here can lead to further improvements in our approach and further reductions in our accident statistics.

THE TEN BASIC PRINCIPLES OF SAFETY

Principle 1

An unsafe act, an unsafe condition, and an accident are all symptoms of something wrong in the management system. We know that many factors contribute to any accident. Our thinking, however, has always suggested that we select one of these as the "proximate" cause of the accident or that we select one unsafe act and/or one unsafe condition. Then we remove that condition or act.

The theory of multiple causation suggests, however, that we trace all the contributing factors to determine their underlying causes.

Principle 2

We can predict that certain sets of circumstances will produce severe injuries. These circumstances can be identified and controlled. This principle states that we can predict severity of accidents under certain conditions and thus turn our attention to severity per se instead of merely hoping to reduce it by attacking frequency.

Statistics show that we have been only partially successful in reducing severity by trying to control frequency. National Safety Council figures show an 80 percent reduction in the frequency rate over the last 40 years. The same source shows that during this period there has been only a 72 percent reduction in the severity rate, a 67 percent reduction in the fatal and permanent total rate, and a 63 percent reduction in the permanent partial disability rate. In recent years, minor injuries have plateaued, getting slightly worse, while days lost (severity) have skyrocketed.

A number of recent studies suggest that severe injuries are fairly predictable in certain situations. Some of these situations involve:

- *Unusual, nonroutine work.* This includes the job that pops up only occasionally and the one-of-a-kind solution. Nonroutine work may arise in both production and nonproduction departments. The normal controls that apply to routine work have little effect in the nonroutine situation.
- *Nonproduction activities.* Much of our safety effort has been directed to production work. However, there is a tremendous potential exposure to loss associated with nonproduction activities such as maintenance and research and development. In these types of activities most work tends to be nonroutine. Because it is nonproduction work, it often does not get much attention with respect to safety, and usually the work is not carried out according to standardized procedures. Severity is predictable here.
- *Sources of high energy.* High energy sources usually can be associated with severity. Electricity, steam, compressed gases, and flammable liquids are examples.
- *Certain construction situations.* Included are high-rise erection, tunneling, and working over water. (Actually, construction severity is an amalgam of the previously described high-severity situations.)
- *Many lifting situations.* Since the 1960s, back strain has been our greatest problem in both frequency and cost.
- *Repetitive motion situations.* In the early 1980s industry began to experience numerous tendonitis and carpal tunnel syndrome claims, many resulting in surgery and considerable lost time.
- *Psychological stress situations.* The mid-1980s noted the emergence of a whole new safety problem—injuries, illnesses, and claims resulting from employees being exposed to stressful environments. As these situations often result in much lost time, and in long term, even permanent disability, they tend to be very costly.
- *Exposure to toxic materials.* Since Bhopal, toxic materials have been recognized as a contributing factor to serious long-term liabilities.

In recent years, severity has taken on a whole new meaning. Incidents such as the chemical fumes escaping at Bhopal, the nuclear disaster at Chernobyl, the shuttle disaster of 1986, and chemical plant explosions have focused unprecedented attention on severe incidents.

Principle 3

Safety should be managed like any other company function. Management should direct the safety effort by setting achievable goals and by planning, organizing, and controlling to achieve them. Perhaps this principle is more important than all the rest. It restates the thought that safety is analogous with quality, cost, and quantity of production. It also goes further and brings the management function into safety (or, rather, safety into the management function). The management function by definition should include safety, but in practice it has not done so. Management has too often shirked its responsibility here, has not led the way, and at best has given "support."

We in the safety profession are often partly at fault. We have not made management lead the way—we only asked for (or hoped for) management support. We have not demonstrated that safety is a management responsibility requiring goal setting, proper planning, good organization, and effective management-oriented controls. At times, we have not even spoken management's language. It will be only when management manages safety as it does other functions that we will see results that reverse our present trends.

Inherent in this principle is the fact that safety is and must be a line function. As management directs the effort by goal setting, planning, organizing, and controlling, it assigns responsibility to line managers and grants them authority to accomplish results. The word "line" here refers not only to first-level supervisors but also to all management-level supervisors above those on the first level, up to the top.

Principle 4

The key to effective line safety performance is management procedures that fix accountability. Any line manager will achieve results in those areas in which he or she is being measured by management. The concept of "accountability" is important for this measurement, and the lack of procedures for fixing accountability is safety's greatest failing. We in safety have preached line responsibility for many years. If we had spent this time devising measurements for fixing accountability of line management, we would still be achieving a reduction in our accident record.

A person who is held accountable will accept the given responsibility. In most cases, someone who is not held accountable will not accept responsibility—he or she will devote the most attention to the things that management is measuring:

production, quality, cost, or any other area in which management is currently exerting pressure.

This principle is extremely important for the implementation of principle 3. Principle 4, in effect, makes principle 3 work.

Principle 5

The function of safety is to locate and define the operational errors that allow accidents to occur. This function can be carried out in two ways: (1) by asking why accidents happen—searching for their root causes; and (2) by asking whether certain known effective controls are being utilized.

Pope and Cresswell suggested that to accomplish our purposes, we in safety would do well to search out not what is wrong with people but what is wrong with the management system that allows accidents to occur.

The role of the safety professional was examined in Chapter 1.

Principle 6

The causes of unsafe behavior can be identified and classified. Some of the classifications are overload (the improper matching of a person's capacity with the load), traps, and the worker's decision to err. Each cause is one that can be controlled.

Principle 6 summarizes the causation model. It suggests that management's task with respect to safety is to identify and deal with the causes of unsafe behavior, not the behavior itself. The model suggests that there are many actions that managers can take to reduce the likelihood of the unsafe act, beyond the two (or three) suggested originally by Heinrich. We simply are not limited to education and enforcement. Actually, we know today that education and enforcement are the two *least* effective things that we can do to improve worker behavior.

An always present aspect and cause of an incident or an accident is human error, which, to restate, results from one or a combination of three things: (1) overload, which is defined as a mismatch between worker capacity and the load placed on the worker in a given state; (2) a decision to err; and (3) traps that are left for the worker in the workplace.

Principle 7

In most cases, unsafe behavior is normal human behavior; it is the result of normal people reacting to their environment. Management's job is to change the environment that leads to the unsafe behavior.

Principle 7 is an extension of principle 6. It suggests that when people act unsafely, they are not dumb, are not careless, are not children who need to be corrected and changed to make them "right." Rather, it suggests that unsafe behavior is the result of an environment that has been constructed by management. In that environment, it is completely logical and normal to act unsafely.

Principle 8

There are three major subsystems that must be dealt with in building an effective safety system: (1) the physical, (2) the managerial, and (3) the behavioral.

Principle 7 reemphasizes that our task is to change the physical and psychological environment that leads people to unsafe behavior.

The role of safety was defined by four elements: (1) analysis, (2) developing systems of control, (3) communicating those systems to the line organization implementing them, and (4) monitoring the results achieved. Principle 8 suggests these four must include systems in all three areas: (1) physical condition control, (2) the management system, and (2) the behavioral environment.

Principle 9

The safety system should fit the culture of the organization. With changing times, the way we manage has changed markedly. And the way we manage safety also must change to be consistent with other functions. To strive for an open and participative culture in an organization and then use a safety program that is directive and authoritarian simply does not work.

Principle 10

There is no one right way to achieve safety in an organization; however, for a safety system to be effective, it must meet certain criteria. The system must:

- *Force supervisory performance.*
- *Involve middle management.*
- *Have top management visibly showing their commitment.*
- *Have employee participation.*
- *Be flexible.*
- *Be perceived as positive.*

The criteria listed in principle 10 are the author's criteria, and the ten principles also are the author's. Keep in mind, however, that the author's belief in these principles and criteria does not make them right for everyone. Each of us must develop what we think are the right principles and criteria. That is our job as safety professionals.

Chapter *13*

Defining and Implementing Change

At this point you have completed two important steps of the three-step change process:

1. You have identified where you are now.
2. You have identified where you want to be.

The rest—providing the difference—is easy. Or is it? Maybe you will have to show some top executives the results of your analysis (where you are now), and maybe you will have to get consensus on where you want to be. Once you have done this, step 3 may or may not be obvious.

This chapter presents some thoughts on how to get to step 3.

Chapter 11 discussed how to interpret the data collected on strengths and weaknesses. Here we start the process of fixing what have been shown to be problem areas. Because individual organizations will have different problems unearthed, we cannot discuss how to fix everything; so in this chapter we will deal with examples—what has been done in different companies.

In Chapter 11 we used an overall chart from one company to illustrate how to diagnose from perception survey results. We will use the same company for further illustration.

From the survey results (see Figure 13-1), these were the organization's primary problems:

1. Lack of employee recognition (47% unfavorable employee response).
2. Lack of supervisory performance (low ratings in Quality of Supervision, Supervisor Training, and Inspections).

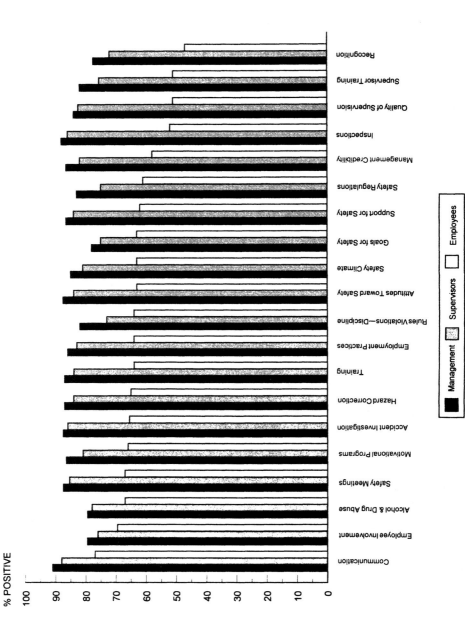

Defining and Implementing Change   *195*

3. Lack of management credibility in safety (low ratings in Management Credibility and Support for Safety).
4. Safety regulations.

Four areas were chosen to be targets for improvement in this company, in this order:

1. Improve supervisory performance.
2. Improve management credibility.
3. Improve employee recognition.
4. Assess safety regulation changes needed.

The next step following analysis of the perception survey results was to look closely at the organization to understand better the reasons behind the low scores.

FIXING PROBLEMS: AN EXAMPLE

Supervisory performance in safety, or in any other area, is determined by a system that holds the supervisor accountable for a defined performance. To ensure supervisory performance, the following requirements must be met:

1. The performance must be defined: what specifically is the supervisor supposed to do?
2. The supervisor must be able to do the defined tasks (be trained).
3. The system (or the boss) must regularly measure the supervisor with a measure that is valid—that tells whether or not the tasks were in fact done, and at an acceptable quality level.
4. The supervisor must think that the tasks are worth doing (be rewarded for performing).

A quick look at the organization's accountability system showed some glaring weaknesses:

1. Performances in safety were not crisply defined; they were fuzzy. There was a management statement of policy that said the supervisor was the key person—was responsible for the safety of the workers. Period.
2. There had been training, but it had taken place years ago; and the training was general, stressing the importance of safety and not defining the tasks a supervisor should do.
3. The supervisor's safety performance was measured monthly by the acci-

dent frequency rate of the department, and at that size unit it was only measuring luck.
4. The reward for a supervisor was a safety award at the end of the year for a perfect safety record.

Needless to say, under the above circumstances supervisors spent their time on other, more important things, such as production and quality where the measures were crisp and valid, and where rewards were clearly present—in the daily numbers game, in the performance appraisal system, and so on.

Therefore, the first fix was to improve supervisory performance.

FIX #1—Define supervisory tasks.
    Measure supervisory performance with a valid (activity) measure.
    Build safety performance into *the daily numbers game* and the company performance appraisal system.

The second major fix involved the perception of management's credibility. Interviews with a number of hourly workers began to shed some light on the reason for low scores here. Over and over again the workers cited three things:

1. Top management was never seen on the shop floor.
2. Top management's messages were never about safety.
3. A number of financial decisions had been made over the years to put money into improving productivity—rather than on safety items that clearly needed fixing.

It appeared to the work force that safety had a low priority with the top people in the company. An interview with top management, however, indicated the opposite:

1. They were genuinely interested in work safety.
2. They did not realize that their visibility was a factor.
3. In choosing how to spend money in the plant, no safety alternative had ever been presented to them.
4. They simply had not realized that they were expected to play a larger role in safety than they had had.

So the second fix involved top management.

> FIX #2—Sit down with top management and help them define a larger, more visible safety role.

The third problem was the need to improve employee recognition. Again, interviewing a number of employees defined the problem clearly. The company had thought of recognition in terms of safety plaques, steak dinners for a good record once a year, or an annual trinket (cap, jacket, etc.). The employees thought of recognition in terms of what happened every day on the job, whether or not the boss ever gave the workers compliments for a job well done. The company norm was to manage by exception, to pay attention only to the things that went wrong—mistakes, errors, goofs, unsafe behavior, and so forth. Thus a third fix was suggested.

> FIX #3—Build a system that assures that management down to the supervisor observes behavior daily and gives positive recognition daily.

The final area targeted for attention was that of the low-rated safety regulations. Analysis of that category had shown that the problem was really one of perception of priorities rather than too many or poor rules, so Fix #2 would already begin to deal with the problem. But for this area to improve, probably several more things might be indicated, such as a fresh look at the rule book to make sure it still made sense, and a system to better prioritize the way management was making decisions on where to put their capital improvements. Hence the fourth and fifth fixes were developed.

> FIX # 4—Reexamine the rules and regulations to sort out the "musts" from things less crucial.

> FIX #5—Define a system to prioritize capital expenditures with safety carrying a higher priority.

The five fixes or recommendations were submitted to top management, and in one short meeting it was decided that:

1. A task force of five line managers would be appointed to clarify and define line managers' and supervisors' safety tasks, measures, and how safety would fit into the daily numbers game and the performance appraisal system.
   - To be completed in one month with a report to the top.
2. Top management, with staff safety help, would define their own visible safety tasks.
   - Report to be completed in two weeks.
3. A second task force of five different line managers and hourly employees would be formed to define the method and the system of assuring that positive recognition would take place, including the training needed and the measurement system.
   - Report to be completed in one month.
4. A third task force of five managers and hourly employees would be formed to revise the rules.
   - Report to be completed in three months.
5. A fourth task force of line managers, maintenance people, and hourly employees would be formed to develop a priority matrix.
   - Report to be completed in one month.
6. All of the above would be reported to all employees in a meeting(s) chaired by the CEO within a week, with a second meeting to be held when the task force reports were in.

The above example showed one company's approach. Analysis of data and determination of the fixes needed were done by management. This often is the way things happen. The task can be done either in staff or in a line organization with a task force, or both. If this is the choice, the fault trees provided in Part II might help. There are, of course, other ways to analyze and to determine action plans by utilizing the hourly workers.

Involvement Approaches

Total Quality Management (TQM) has been very popular in industry in recent years. It goes by different names and acronyms, but the process seems to be fairly consistent in application, consisting of:

1. Building a better culture.
2. Using hourly employees in problem solving and decision making.
3. Using some tools called Statistical Process Control (SPC) tools.

Defining and Implementing Change

SPC is a popular concept in the United States today, but has not always been. U.S. industry basically spurned the whole idea of SPC in the 1950s, whereas today it wholeheartedly endorses it. But as we disregarded it in the 1950s, Japanese industry bought the concept, implemented the tools, and turned itself completely around. By ignoring it, U.S. industry almost lost its world industrial leadership position.

Is SPC a powerful tool? When the United States failed to adopt it, U.S. productivity dropped 15 percent between 1960 and 1980. Having bought it, the Japanese increased their productivity by 150 percent in the same period—and their SPC thrusts were for *quality,* not productivity.

In the 1980s, American managers rushed to Japan to learn its secret, only to find that the secret was what the U.S. managers had rejected 20 years earlier.

The main man behind SPC was Dr. W. Edwards Deming. Born in 1900 in Sioux City, Iowa, and raised in Cody and Powell, Wyoming, he studied at the University of Wyoming in mathematics, graduating in 1921; at the University of Colorado for a master's degree in mathematics and physics; and at Yale University for a doctorate in physics. He earned his way through school with summer employment at Western Electric Company, at the Hawthorne Works in Cicero, Illinois. (*Note:* This was during the time of the famous "Hawthorne Studies.") His familiarity with the Hawthorne Studies seems to have been a major factor in his approaches in later years.

What is the Deming Management Method? Basically it is a two-pronged approach: using statistical tools for continuous process improvement, and using the involvement of all employees in the organization.

The use of statistical methods is not an all-purpose remedy for every corporate problem. But it is a rational, logical, and organized way to create a system that can assure continuing, ongoing improvements in quality, productivity, and safety simultaneously.

The statistical approach, as advocated by Dr. Deming, fits well with the changing management control concept in many American companies, moving from a concept of "detection" to one of "prevention." The drawback associated with "detection" is that an unacceptable product or an accident must be produced or occur before people can determine how to adjust the process. Obviously, this wastes resources, for it costs just as much to produce an unacceptable product as an acceptable one, yet the unacceptable product must still be either reworked or scrapped. The ideal situation would involve being able to monitor (and, if necessary, adjust) the process on a periodic, "real-time" basis in order to minimize the possibility of producing unacceptable products or accidents. This approach might be called "prevention." If we can (and indeed we can) monitor before the fact of an accident, and adjust the process (or the behavior) before someone is hurt, we are much ahead in every way.

Statistical techniques are among the best tools for evaluating the selective measurements this approach requires. They provide a method for logically and systematically evaluating information. Specifically, they help determine process stability, enable producers to consistently meet consumer requirements, and point to the causes of accidents and other problems (should they arise).

The two parts of the Deming Management Method may be briefly described as follows:

1. *SPC tools:* The Deming approach offers a number of tools to continually improve the process. But the tools are nothing more than a way to achieve the central message: "Assess or fix the process continually; the results then take care of themselves." In other words, assess the process; measure the process; analyze the process; fix the process; etc. Concentrate totally on the process and forget the bottom line because that is a result of the process, and will only be a reflection of the process.
2. *Employee involvement:* Probably learned at the Hawthorne Works, the second part of the Deming approach suggests that continuous process improvement can come only with employee participation and ownership.

Both approaches are included in the Deming philosophy. This philosophy (or culture, if you prefer) is summarized by Deming's 14 points:

1. Create constancy of purpose for improvement of product and service.
2. Adopt the new philosophy; we are in a new economic age.
3. Cease dependence on inspection as a way to achieve quality.
4. End the practice of awarding business on the basis of price tag.
5. Constantly and forever improve the system of production and service; the system includes people.
6. Institute training on the job.
7. Institute improved supervision.
8. Drive out fear.
9. Break down barriers between departments.
10. Eliminate slogans and targets asking for increased productivity without providing methods.
11. Eliminate numerical quotas.
12. Remove barriers that stand between the hourly worker and his right to pride of workmanship . . . [and do] the same for all salaried people.
13. Institute a vigorous program of education and retraining.
14. Put everybody in the company to work to accomplish the transformation.

Deming's philosophy, or these 14 points, speak to quality, and to a general management culture. We believe they also speak to safety.

The Corporate Safety Obligations

I have taken Deming's 14 points and summarized, combined, and changed them into my perception of a safety philosophy that fits the Deming philosophy. They look like this:

1. Concentrate on the long-range goal of having a world-class system in place, not short-term annual accident goals.
2. Discard the old philosophy of accepting accidents—they are not acceptable.
3. Use statistical techniques to identify the two sources of accidents—the system and human error.
4. Institute more thorough job training.
5. Eliminate the dependence on accident investigation to unearth flaws in the system—use proactive approaches, sampling of behavior, fishbone diagrams, flow charts, and so on, to achieve continuous system improvement.
6. Provide supervision (and employees) with knowledge of statistical methods (sampling, control charts, etc.) and ensure that they are used to identify defects for further study.
7. Reduce fear throughout the organization by encouraging all employees to point out system defects, and to help solve defects.
8. Help reduce accidents by designing safety into the process. Train research and design people on safety.
9. Eliminate the use of slogans, incentives, posters, and gimmicks to encourage safety.
10. Closely examine work standards to remove accident traps.

The ten points define a "Safety Culture," or a culture where safety is shared as a key value of the organization.

You will note a marked departure from traditional safety thinking:

- Safety not measured by accident statistics.
- Safety becoming a system, not a program.
- Use of statistical techniques for improvement.
- Reduction or elimination of accident investigations.
- Use of safety sampling and SPC tools.
- Elimination of blame for "unsafe acts."
- Focus on system improvement and causes of human error.
- Encouragement of "whistle-blowers."
- Employee involvement in problem solving and decision making.
- Using ergonomics in design.

- No slogans or gimmicks.
- Removal of traps, of system-caused human errors.

Safety professionals have used statistics and statistical concepts such as safety sampling, safe-T-scores, and control charts for many years, but these and other techniques have never become an integral part of most safety systems.

The SPC Tools

The following section describes the basic tools to use in statistical process control.

The Problem-Solving Tools

Histograms and Paretos are probably used by safety professionals today more than any other SPC tools.

Histograms. A histogram is a bar graph displaying a frequency distribution (Figure 13-2).

Paretos. A Pareto is a bar graph showing results in descending order of magnitude (Figure 13-3). Paretos can be used again and again, analyzing further and further what the problem is.

Here is a real-life example of the use of Paretos, as well as other tools:

The top executive of the XYZ Corporation made a decision to improve the safety results in his organization. Spurred by a fatality and the realization that his

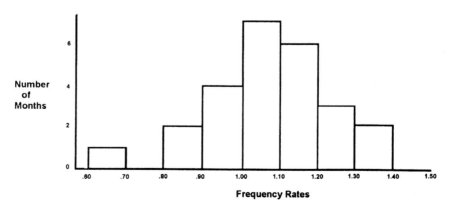

Figure 13-2. A histogram.

Defining and Implementing Change

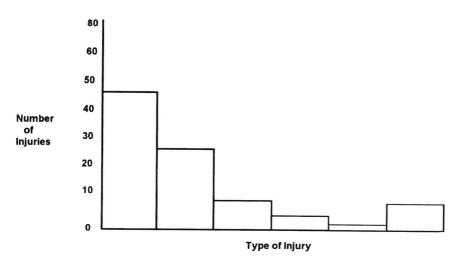

Figure 13-3. A Pareto.

organization was one of the worst in the industry, he did what executives typically do: he put pressure on employees to "turn it around."

To start the process, an employee group was formed, and it first used a series of Pareto charts to find out where the biggest problems were:

- Pareto 1: Where is the problem? A Pareto of the fifteen plants in the corporation showed that the problem was primarily in five plants. A quick computation showed that bringing the accident record down in those five plants to the level of the other ten would make the company second best in the industry. An additional halving of that figure would make it the best.
- Pareto 2 answered which type of injury was predominant in those five plants. The problem was soft tissue injury in wrists, shoulders, and backs.
- Pareto 3 showed which department caused most of the soft tissue injuries.

From the three Pareto charts it was decided that a step change could be made through reducing the soft tissue injuries in two departments of five plants, involving the packers in the packaging department and the box handlers in shipping.

Fishbone Diagrams. A fishbone is a cause-and-effect diagram for analyzing problems.

Continuing our example, a fishbone diagram (Figure 13-4) was next used to determine the causes of the soft tissue injuries in each of the above departments.

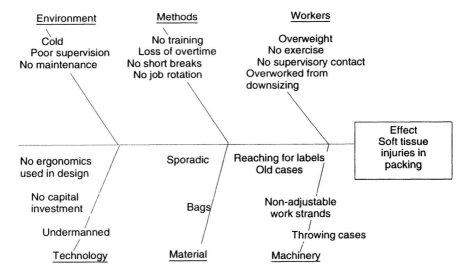

Figure 13-4. Fishbone diagram used in text example.

To get the best possible input here, the workers themselves were asked to problem-solve using the fishbone approach.

Kaoru Ishikawa, whose *Guide to Quality Control* was written for Japanese workers and is now the most widely read book on basic statistics for quality in the United States, outlines these benefits from cause-and-effect diagrams:

1. The creation process itself is educational. It gets a discussion going, and people learn from each other.
2. It helps a group focus on the issue at hand, reducing complaints and irrelevant discussion.
3. It results in an active search for the cause.
4. Data often must be collected.
5. It demonstrates the level of understanding. The more complex the diagram, the more sophisticated the workers are about the process.
6. It can be used for any problem.

Finally, the group in our example decided which factors to concentrate on:

- Get adjustable packing tables.
- Do preshift warm-up exercises.
- Use wrist bands.
- Rotate packers hourly.

Defining and Implementing Change   205

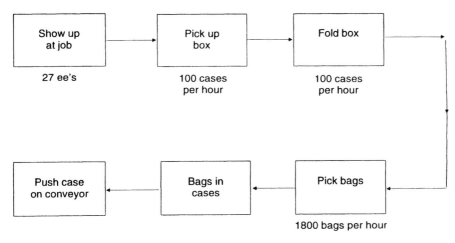

Figure 13-5.   Flow chart used in text example.

A substantial improvement was achieved. The approach became a normal way of life in the five plants and was used on other safety, quality, and productivity problems. In one year the five worst plants became the five best plants, and the company became the best in its industry. The secret here is the same secret that Dr. Deming brought to Japan in the 1950s: some SPC concepts coupled with employee involvement in the problem-solving process.

Flow Charts.   In the example flow charts were also used (see Figure 13-5). Flow charts can be constructed for anything, just like fishbones.

Scatter Diagrams.   These are graphs depicting the correlation of two characteristics; a scatter diagram allows one to see if there is a relationship between the characteristics. For instance, suppose that you are concerned that your audit may not be telling you how effective your safety program really is. A simple way to test your audit is with a scatter diagram: simply plot your audit scores with the accident results for the year for all of your locations. (See Figure 13-6.)

What you end up with is a picture showing whether or not there is a correlation. If your graph looks like a straight line, you can trust your audit; it is a good indicator of the effectiveness of your program. If it looks like a circle, you have no correlation.

Keep in mind that the figure has to face the right way. In this case we want a negative correlation; high audit scores mean low accident frequency.

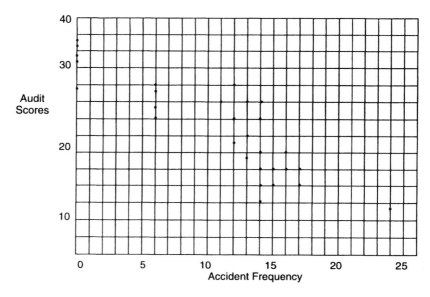

Figure 13-6. A scatter diagram.

The Control Tools

In addition to the problem-solving tools, SPC also offers control tools: run charts and control charts.

Run Charts. These charts provide a description of what is happening over time. They can take the form of bar charts or line charts.

Control Charts. Of the analytical tools described above, the control chart is a particularly useful and effective device (Figure 13-7). On the chart the observed accident rates are plotted against time, producing an overall accident rate or mean for the entire period. Finally, upper and lower control limits are computed so that the probability of an accident rate exceeding the limits by chance alone is very small. The control chart is based upon a bell-shaped curve, which is applicable to many things, such as height and weight of people, frequency of accidents, and so on.

From a study of statistical control chart data, it can be determined whether the system is a relatively constant one. If so, there is a stable situation. Conversely, if the accident rate exceeds the upper limit, that signals a change for which there is an assignable cause. Similarly, when a point falls below the lower limit, there has been a significant change for the better.

Statistical control techniques offer a means for making the work of accident reduction more effective and efficient. They cannot assign cause, but they can point out where and when to look for causes.

Defining and Implementing Change

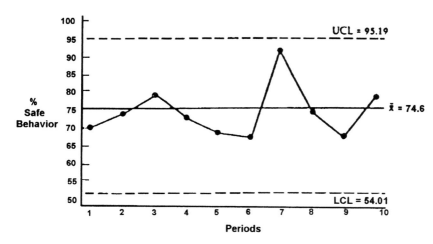

Figure 13-7. Control chart.

Input Data

All of the SPC tools can be used with various kinds of data as input. They are very valuable, as measurement always has been and still is safety's largest problem.

Historically, we have concentrated on accident statistics as our measuring sticks; and we have used them pretty much for all measurement purposes: to compare, to diagnose, to motivate, and so on. This has been disastrous, for the accident statistics usually are invalid measures of safety system success; they almost never diagnose our problems; and in small units they measure only luck.

We have a number of measures that we can choose, and they come in two categories: (1) activity (or performance) measures and (2) results measures. As a general rule, in selecting measuring devices, one should use only activity measures at the lower managerial levels, and primarily activity measures (with some results measures) at the middle–upper management levels, reserving the pure results measures for the executive level. With rare exceptions (such as safety sampling), this rule of thumb gives us statistical validity and is extremely important.

Keep in mind the central Deming philosophy that we should be dealing with data that tell us how the process is doing; and usually accident data do not do that, whereas sampling and performance data do.

The Deming approach has been proved; it works. Remember its two-pronged approach: using SPC tools and using employee involvement. Often our focus is only on the tools, but keep in mind that is only half of the picture. Employee participation is equally important.

The tools work, and they work in safety. They are used for problem solving and to monitor—to monitor the continuous improvement of the safety system in the organization.

The key to everything is involvement of the hourly workers. As they assist in the process of fixing the areas that are weak, they have a feeling of ownership; get feelings of achievement, responsibility, and recognition (the motivators); and can see a new culture in the organization that says they are important, and safety is a key value.

Chapter *14*

Sources of Help

This chapter looks at sources of help available to you in analysis and improvement of the behavioral system, safety system, and physical environment. Probably the first point to be made is that the task of analysis and improvement is entirely yours, and cannot be delegated to the outside. There are people and agencies that can help, but the responsibility cannot be delegated. Second, the analysis is better if made internally because you know your organization better than any outsider. There are, however, times and circumstances when an outsider's help can be invaluable.

WHEN TO SEEK OUTSIDE HELP

Kenneth Albert suggests that there are six specific times when outside help should be obtained:

1. When special expertise is essential.
2. For a politically sensitive issue.
3. When impartiality is necessary.
4. If time is critical.
5. If anonymity must be maintained.
6. When the prestige of an outsider would be helpful.

Although Albert is speaking in a nonsafety vein, some of the above points seem valid in determining the need for an outside safety consultant. If there is no one in your organization who is familiar with OSHA, it might be helpful to call in an OSHA expert to assist. Perhaps, more realistically, if there is no one in your organizational structure able to analyze the behavioral system, it might serve you well to locate such a person.

Often a safety problem is intertwined with managerial personalities and is extremely difficult to solve internally. Here an outsider might well offer a solution acceptable to top management. If you need an analysis in a hurry with some suggested approaches to a solution, it often can be done faster by an outside consultant, and often an outsider's recommendation will seem to carry more weight than the insider's.

In the field of safety, it appears as though most organizations need outside help with the behavioral system analysis, many with the safety system analysis, and a few with the physical condition analysis. However, in looking at what is available among safety consultants, the exact opposite is true. Physical condition consultants abound everywhere, safety system analysts are quite rare, and behavioral analysis experts are almost nonexistent.

WHAT EXTERNAL HELP IS AVAILABLE

External help might be classified into these categories:

1. Insurance company field safety engineers or consultants.
2. Government safety consultants under OSHA direction.
3. Private consulting firms.
4. Part-time private consultants.
5. National Safety Council consultants.
6. Other association consultants.

We look at these consultants in some detail in this section, including a discussion on "Locating the Consultant." There are no doubt other consultants who might not fit any of the above categories, but generally most will fit into them.

Insurance Consultants

Most U.S. consultants fit into this category. Insurance consultants comprise a large percentage of the safety engineers in the United States, and many others (if not most) get their safety start in insurance companies. Almost all companies, except the very large and self-insured, are helped routinely by insurance loss-control representatives.

Government Consultants

Who provides the consulting service varies from state to state. In some cases, the state contracts with the federal OSHA to do the consulting; in others, private organizations get the contract; and in some, the state university will provide the consulting service to the state's employers who wish it. Utilizing government consultants is completely voluntary. They will not visit any place of business unless invited. Before choosing to invite them, however, any organization ought to find out a little about the voluntary compliance program.

The stated goal of the on-site consultation program by OSHA is to obtain "safety and [a] healthful workplace for employees." Thus by definition the consulting will pertain only to physical conditions. If this is where you need help, this route might be possible. If, however, you need consulting help with the safety system or the behavioral system, this is the wrong place to go.

The defined on-site responsibilities of the OSHA consultant are:

1. To identify and properly classify hazards.
2. To recommend corrective measures (short of engineering assistance).
3. To arrange abatement dates for serious hazards.
4. To report to his or her supervisor any employer-unabated serious hazards.
5. To follow up on employer actions.

You might note here that there are some aspects of receiving consulting service by this route that are unusual. Their purpose is to help improve physical conditions, but in two instances the consultants have additional duties:

1. In the case of serious violations of the OSHA standards, they must set abatement dates and follow up.
2. In the case of imminent violations of the OSHA standards, they must refer them either to their supervisor and on up the Bureau of Labor hierarchy or to the compliance staff for your immediate compliance.

Private Consulting Firms

A third source of external help is the private consultant (full-time) or the private consulting firms. They can be called in to consult in any area you want—behavioral, safety system, or physical—and with none of the strings mentioned above in using OSHA consultants. The only difficulty is ensuring that you have chosen wisely and obtained a consultant with the necessary skills and knowledge to do what you want done.

Part-Time Private Consultants

The fourth place to locate a private consultant is among those individuals who consult on a part-time basis to supplement their incomes. This person usually is either a retired safety professional who would like to keep active or a college or university professor who needs to supplement his or her income and hopes to keep knowledgeable about the real world.

Here again the problem is locating these people and ensuring that you have hired a person with the competencies that you need. The simple fact that a person has 40 years in safety, or teaches safety, does not mean that his or her competencies are ensured.

Other Consultants

An additional source in recent years is consultants available from the National Safety Council, and others.

Locating the Consultant

In the first two categories above, this is easy; just call your worker's compensation carrier or the local OSHA office and ask for someone to come to your organization. The next two categories present more difficulties.

Several organizations put out directories of consultants. The American Society of Safety Engineers publishes a national directory, which included at last count nearly 300 names. In using this to help select a consultant, there seem to be some real problems. An analysis of the names showed that about half were individuals who indicated that they were for hire and who had no connection stated. Some of these are retired; some work for companies and want some additional side income; some are full-time safety consultants. You cannot tell the full-time ones from the moonlighters. An additional 32 percent were identified as being connected with a consulting firm, 5 percent were connected with a university, 3 percent were insurance brokers, 3 percent were connected with manufacturing companies, and 1 percent were with a state government. Actually, this directory, while advertised as a document telling "where the occupational safety/health experts are," is really a roster of those people who have paid dues and are members of the consultant's division of the ASSE.

There is no easy answer to finding the consultant who has the expertise you might need. Probably the best approach is the professional organization, the ASSE, and the document indicated above, keeping in mind the above comments. Some additional thoughts on choosing and analyzing what you get from consultants are included later in the chapter.

First we look at the most available of the outside consultants.

Sources of Help

Insurance Consulting

Since the beginning of the industrial safety movement in the early 1900s, the insurance industry has been in the middle of the safety effort. Until a few years ago, the only possible external help for most companies had been the help available from the insurance carrier. While this is no longer true, the insurance consultant is still far and away the most used. Estimates of the percentage of safety professionals in the insurance industry have ranged from 33 to 70 percent. About 29 percent of the membership of the professional organization (ASSE) are in the insurance industry. Many safety professionals, including the author, grew up in that industry.

The Services Insurance Carriers Offer. The safety services department of any large insurance company is charged with three specific functions (see Figure 14-1):

1. A sales assist function.
2. An underwriting assist function.
3. A customer service function.

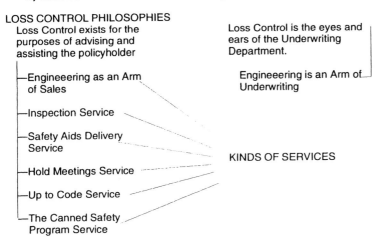

Figure 14-1. Safety services department functions.

Only the third of these is of value to the customer. Historically, most companies started with a primary emphasis on function 2, the underwriting assist function, where the field representative is the "eyes and ears" of the underwriting department. This field job is one of observing and reporting what is going on at the policyholder's place of business back to the desk-bound underwriter, who can then, based upon the information, determine desirability and pricing of the insurance. The third function, that of customer service, seems to have evolved a little later, and never in some organizations. This function consists of doing something for the customer that improves his or her safety program and reduces the likelihood of that customer's having accidents and financial loss. What that something is varies considerably from company to company. The first function, that of helping to sell insurance to more customers, is a by-product of the service function, convincing the potential customer that the insurer can provide some help.

As a result of past experience, different philosophies have emerged that dictate the value of the service the insurance company is able to provide. In some companies the safety services department is still very much a part of the underwriting function. Its duties are to observe and report. In some companies the engineering department reports into the underwriting department. In other, more modern insurance companies the loss-control department is independent, existing primarily to serve the customer and secondarily to sell and to assist the underwriter. The two conflicting philosophies are depicted in Figure 14-1.

If the loss-control department is not an arm of underwriting, it has a chance of being able to provide good service to a customer. It also may be quite ineffective, as there are all kinds of ways to make sure the service provided is not effective. Some of these are listed in Figure 14-1 also. When the service has as its primary mission to assist sales, the customer service will suffer. When the service is an inspection-only service, as is very prevalent, the service suffers because the safety system and the behavioral system are totally overlooked. When the service consists of the delivery of safety aids and materials and nothing else, it is a meaningless service. When the service consists primarily or totally of merely holding safety meetings for a customer or merely of ensuring that physical conditions are up to code, it is a weak service. Perhaps the least effective service (although it is fairly common) consists of the carrier's representative bringing out the canned safety program that the carrier's home office has devised for use at all insured companies.

Depending upon what philosophies are behind the service of the carrier, there may have developed additional services beyond the service of the representative who calls on the customer. Figure 14-2 outlines some typical additional services. Industrial hygiene services can be particularly useful to customers. Nursing and specialist services can also be helpful, depending upon current needs. Training services are somewhat less common but are perhaps the most valuable kind of service that carriers have. Materials are available from all, at least providing an alternative to purchasing them from the National Safety Council or some other source.

Sources of Help

Safety engineers service	Training
Inspection	Safety management
Engineering	Supervisory training
Consulting	Speciality courses
Industrial hygiene services	Fleet safety
Dust	Products safety
Fumes	Construction
Gases	Driver training
Vapors	Classroom
Noise	In plant
Nursing services	Behind the wheel
Occupational health	Programmed instruction
Rehabilitation	Symposia
Specialists	Materials
Fleet	Forms
Construction	Literature
Products	Posters
Radiation	Periodicals
Electricity	Films
Elevator	Awards
Fire	
Press	
Medical	
Stress	

Figure 14-2. Loss-control services.

Government Consultants

As with the insurance consultants, there are additional considerations in determining whether or not to avail yourself of this service from the government. Some of those might be:

1. Whether the referral prospects are of concern to you.
2. The competence of the people.
3. The limited scope of the consulting.
4. The inability to direct the consulting focus.

Probably the first consideration is whether or not you wish to become involved with the possibility of a referral, as mentioned earlier. With other kinds of consultants (either insurance or private), whatever is found and discussed is strictly between you and the consultant. Whatever you wish to do about what is found is totally up to you. You retain control. With OSHA-funded consultants this is not totally true. With two kinds of hazards, it may no longer be between just you and the consultants, and you may not be able to retain the decision power over what to do about the hazard and when to do it. The hazards that will be pointed out will be law violations.

The final consideration might be whether or not you wish to control the focus of the consulting help or not. With OSHA consulting, you no doubt will lose control. If you want help in one specific area, it is quite possible that the help you get will not be confined to that area. Because you are not the "customer," you cannot direct the energies of these consultants, as you can with any other consultants.

The OSHA consultants can assist in determining whether or not you are in compliance with standards. This is an extremely narrow focus and has many weaknesses, as pointed out by G. Peters in his article "Why Only a Fool Relies on Safety Standards" in *Professional Safety:* "For those who know little about safety, it seems quite plausible and reasonable to expect that the existence of good safety standards and a sufficient conformance to those standards should be an adequate measure of safety assurance." Dr. Peters suggests that not only is this not true, but also that reliance on standards will subvert professional activities that are needed to reduce loss: "The original and best use of standards was to assure the identity of components, so that mass production of goods was possible. Similarly, specifications for products, processes, and services was necessary to obtain sufficient definition that bidding, subcontracting, and resolution of possible contractual disputes was expedited." He then outlines the limitations of safety standards:[1]

1. Outdated Criteria

A [ten-year-old] standard, being applied [now], suffers from the fact that the basic decision-making criteria used in formulating the contents of the standard is, obviously, ten years out-of-date. For example, the criteria utilized to determine the level of acceptable risk may have been valid at the time the standard was published, but those criteria may have been appreciably lowered, in terms of the tolerable risk levels, by the effects of subsequent product liability case law. In fact, the cost–benefit ratios may have been significantly altered because of changes in technical feasibility, better data, new discoveries, different production cost estimates, and the ultimate cost of injuries. So the factors that are evaluated to determine the level of acceptable risk may be quite different ten years after a standard has been promulgated.

2. Inconsistent Requirements

Which standards are right and which are wrong when they each have different requirements. For example, capacitive circuits are to be discharged, upon power shut-off, to less than 50 volts in one minute according to ANSI, CI, and Underwriters Laboratory Standard 170, to 30 volts in 2 seconds according to Military Standard 454, and to 60 volts in 10 seconds according to a major company product safety manual. Which is it: 30, 50, or 60 volts in 2, 10 or 60 seconds?

---

[1]Reprinted with permission from George Peters, Why only a fool relies on safety standards, *Professional Safety,* January 1966.

Obviously, different groups of experts differ in their appraisal of hazards, risks, and dangers, so that their safety requirements and remedies will differ. Therefore, there should be some appraisal of who formulated the standard, what "political" effects may have biased the published requirements, and what specifics may be distinguished between apparently inconsistent requirements.

3. Risk Reduction, Not Hazard Elimination

For any particular hazard identified in a standard, the remedies may be 100% effective, 50% effective, or purely cosmetic or ineffective with no appreciable risk reduction for that hazard. The level of risk remaining after compliance with the provisions of a standard may be a matter of deliberate choice, or it may be the result of an arbitrary uninformed selection. For example, the first federal safety standard, under the authority of the consumer Product Safety Act, dealt with the hazard of shallow water at the end of water slides used at swimming pools. A head first belly slide into shallow water had been identified as a significant cause of severe quadraplegic injuries. After determining the depths of water that would result in varying degrees of injury reduction, the standards committee voted for a 50% risk reduction water depth. Therefore, if a safety engineer complies with that federal safety standard for that particular hazard, he would be accepting the continuance of severe injuries at a rate that was 50% of the historical rate. In other words, full compliance means that you are half-safe. It does not mean that all risks have been completely eliminated nor even reduced to that which are economically and technically feasible.

4. Incomplete Coverage

It is well known that any one safety standard cannot cover all known or predictable hazards. If it did, it might be a book-length fine-print document instead of a concise recipe of key ingredients. Similarly, we cannot have standards that cover every substance, product, process, or service. Should we have standards for all two million chemical compounds that exist today?

Obviously, standards cannot cover all interactions, additive effects, synergisms, and potentiating consequences of compounding, assembly, or use. Thus, we will always have such incomplete coverage by safety standards, that human judgment must predominate on safety issues.

5. Minimal May Be Sub-minimal

To avoid anti-trust problems, the standards development process should be open to all segments of the affected industry. If it somehow disadvantages any particular company, there may be a cry of "foul" and arguments about arbitrary restraint of commerce. This means that the standard has to be written in such a way that almost all products are not excluded from the marketplace. In effect, the standard cannot upgrade what is being produced or it will disadvantage some reluctant company, place it at an economic disadvantage, and raise the specter of anti-trust litigation. In essence, this means that the voluntary consensus standard too often must be primarily ineffective or evadable or discretionary in terms of its safety requirements. Instead of a minimal level of safety, it may die-cast sub-minimal risks of harm. The safety professional should be able to recognize which standards are so disabled and not rely upon them in any significant fashion.

6. One Man's Views

Although standards development committees have been broadened in representation, they may be dominated by one company or one man. If the chairman of the standards development committee is an aggressive officer of a company vitally interested in the standard, then common sense suggests that things will go his way, that no disadvantage will accrue to his company, and that he is representing his company's interests to the very best of his ability. Even for a federal standard, the bulk of the money may be forthcoming from only one manufacturer.

Therefore, regardless of the entity that issues a standard, frequently it reflects only the restricted views of one or more of a small group of people who have some particular motivation or self-interest in pushing-through the standard. In essence, it is nothing more than an expression of opinions of a committee which may or may not be modified by the political compromises of a reviewing authority. It is man-made, not God given. It is fallible, not perfection. It is for guidance, not reliance.

7. Stifles Research

Undue reliance on standards may result in neglect of safety analyses, safety testing, safety research, and quality assurance measures needed for a specific product or process. . . . Undue reliance on standards may lead to safety senility rather than creative solutions to actualized or predictable safety problems.

8. Creates Higher Implied Standards

When the Society of Automotive Engineers published a children's snowmobile standard describing snowmobiles intended for six year old children that were to be placarded as restricted to off-road use, doesn't that suggest that children of limited reading ability, immature judgment, and underdeveloped muscular coordination will, predictably, drive snowmobiles on ice covered public roads? The warning on the snowmobile states that its use on public streets, roads and highways "may be" hazardous. Thus, it recognizes that there is a probability of use of snowmobiles on highways, by children, and it suggests a possibility of safe use on those roadways. The implied safety standard, implicit in such language, is that the manufacturer can design and furnish a snowmobile that is safe for use by six year old children, on public highways, under conditions of ice and snow. Such unwritten and unintended, but implied, safety requirements, demonstrate why the language, content, and substance of any safety standard remains far less than is necessary to properly guide the safety expert. It also graphically demonstrates how abject reliance on every detail of safety standard could create unnecessary safety problems.

9. Legal Obligations Undiminished by Standards

Some people seem to believe that diligence in complying with government standards, rules, and regulations will somehow protect them from lawsuits, if injuries or property damage result. They may hear about government agency requirements, administrative law decisions, and contractual agreements between the government and its contractors, and confuse these topics with the higher safety requirements imposed by the judge-made tort law and what is actually happening in the more numerous and important common law actions in all fifty states. There should be no question, in anyone's mind, that mere compliance with safety standards or the rules of a government agency does not relieve anyone of liability at the basic level of the common law.

10. Improper Allocation of Resources

The quest for appropriate safety standards, in general, has been found to be unbelievably lengthy and costly. Time and money flow like water whether the effort is paid by contract or by allocated budget, or whether it consists of ostensibly free donated voluntary activities. For example, it would seem relatively easy to improve upon several different standards for portable ladders. The ladder standards have been in existence for some time; many people have tested, analyzed, or had personal experience with ladders; for a long time it has been an insurer's loss control headache; and it is a simple product. But, the Consumer Product Safety Commission started that project, about two years and a million dollars ago, and the results are still in question....

In fact, high cost, long delay, limited effectiveness, and self-serving obscuration suggest that it is improper to allocate a major proportion of limited resources to this single aspect of safety. For example, if half the money spent on standards development, by the Consumer Product Safety Commission had been allocated to university research publications or engineering source books on safe product design, the utilitarian resources available to engineers and scientists would have resulted in a quantum leap forward in safety technology useful to those who design and manufacture consumer products.

In conclusion, let me restate my belief that safety standards are an essential ingredient for assured safety. But, too many people are being crippled and killed by excessive reliance on the speculative value of such standards, by improper allocation of limited resources to standards development, and by failure to recognize the true character and use of such standards. Only the innocent would do without standards, but only a fool relies upon such standards in the absence of the professional product and quality assurance efforts that are required in the socio-legal technological climate of the world in which we do business and in which we should enjoy freedom from unreasonable risks of harm.

Dr. Peters is this country's leading authority on product safety. He states the case well against dependence on standards to achieve safety.

Private Consulting

With the private consultants—either consulting firms or individual consultants, full- or part-time—we start out with no built-in drawbacks. The private consultant does not have a required referral system, the relationship is only between you and the consultant, the scope of his or her consulting is limited only by his or her knowledge, and you can very directly control the focus of the help because you are the "customer."

Thus the only thing you have to worry about is whether or not the consultant is competent in the areas where you need help, and whether or not you wish to pay his or her fee.

Figure 14-3 shows some information extracted from the consulting manual *Professional Practices in Management Consulting,* by the Association of Consulting Management Engineers. Here a consultant is defined, and his or her approach is discussed.

G. Lippitt has written extensively on the consulting process. He has identified

---

1. *Diagnose the problem* of the customer and *define it* in terms that lead *to agreement* as to the nature of the problem.
2. Investigate the situation and consider which of various possible approaches will best serve the interests of the customer.
3. Suggest the course of action which will accomplish the maximum benefits for the customer in proportion to the required expenditures of time, effort, and dollars.
4. *Present plans* for carrying out the solution *in a way* that will *secure the understanding, agreement, and cooperation* of the customer.
5. Point out the side benefits that may result from successful accomplishment of the suggestions.
6. Make sure the customer understands clearly what must be done, how long it will probably take, and the approximate cost.

Figure 14-3. Management consultant's approach to service.

Sources of Help

---

1. Helps management examine organizational problems (e.g., organizes a management meeting for problem-identification in the problem relationships between home and field office personnel)
2. Helps management examine the contribution of the proper dialogue to these problems (e.g., in relation to home and field office problems, explores with management how a conference on communication blocks might lead to problem-solving)
3. Helps examine the long- and short-range objectives of the renewal action (e.g., involves management in refining objectives and in setting goals)
4. Explores with management alternatives to renewal plans
5. Develops, with management, the renewal plans (e.g., based on the objectives, works with a task force to develop a process with built-in evaluation rather than simply submit an independently developed plan to management for approval)
6. Explores appropriate resources to implement renewal plan (e.g., provides management with a variety of resources both inside and outside the organization. The renewal stimulator must help management to understand what each resource can contribute to effective problem-solving)
7. Provides consultation for management on evaluation and review of renewal process (e.g., evaluation must be in terms of problem-solving; working with management, the renewal stimulator must assess the current status of the problem, rather than check whether or not certain activities have been conducted)
8. Explores with management the follow-up steps necessary to reinforce problem-solving and outputs from the renewal process (e.g., encourages management to look at the implications of the steps taken so far, and to assess the current status of the organization in terms of other actions that might be necessary to follow up on implementation of the renewal process)

Figure 14-4. Eight specific consultant activities.

eight specific consulting activities (Figure 14-4), and five different consulting approaches (Figure 14-5). He describes the approaches:[2]

> In carrying out a "helping" relationship to management, a renewal stimulator will find himself operating along the continuum of counsulting roles. . . . Here I have illustrated some of the major helping relationships from directive to primarily nondirective consultations.
>
> POSITION 1: ADVISES MANAGEMENT OF A PROPER APPROACH. In certain circumstances management may be attempting to solve a problem by using a medium or method which the renewal stimulator, from his professional experience, feels will not work. He may need to use his best persuasive skills, especially if time is short, to convince management not to use that particular approach.
>
> POSITION 2: GIVES EXPERT ADVICE TO MANAGEMENT. There will be occasions when organizational management will expect the renewal

---

[2]Reprinted from G. Lippitt, *Organizational Renewal: Achieving Viability in a Changing World*, Englewood Cliffs, NJ: Prentice-Hall, 1973.

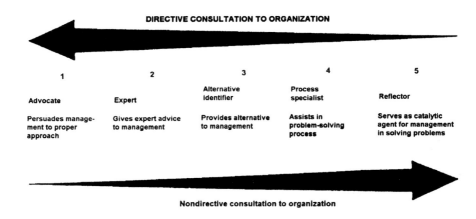

Figure 14-5. Consultant approaches. (From G. Lippitt, *Organizational Renewal,* Englewood Cliffs, NJ: Prentice-Hall, 1973.)

stimulator to answer a technical question; for example, a question about the value of a certain kind of activity and its utilization to the organization in moving toward the next stage of its growth needs.

POSITION 3: PROVIDES ALTERNATIVES TO MANAGEMENT. The renewal stimulator may offer alternatives to management in the solution of a major organization problem. The situation is not one in which he is the implementer of a solution, but one where he recognizes the values of identifying alternatives for management in confronting the learning aspect of a problem and the various implications in organizational functioning.

POSITION 4: ASSISTS IN PROBLEM-SOLVING PROCESS. In this situation, the renewal stimulator serves as a process specialist and consultant to management. He does not get involved in the "content" of the problem, which may be outside his area of competence. Rather, he helps management maintain the quality of its process, problem-solving, and planning through his skill as a specialist in the area of interpersonal relations.

POSITION 5: SERVES AS CATALYST FOR MANAGEMENT PROBLEM-SOLVING. In this last category, the renewal stimulator may only ask questions for management to take into account as it considers a certain direction, action, or policy.

Sources of Help

With the above general thoughts on consulting as background, next we look in detail at how to choose and analyze what you are getting from a consulting firm or consultant.

CHOOSING A CONSULTANT

We might start out by looking at the choosing of a consultant, and suggest a process such as that given by Figure 14-6. Whether or not to use a consultant, and which one to use, ought to be determined by your defined needs and by what kinds of skills and knowledge the consultant would need to be of real help to you. Then, it would seem logical to look (both externally and internally) for individuals or groups that have those skills and knowledge. You may determine as a result of this process to do the job without external help—to locate the needed skills internally and apply those skills to your defined safety problems—or you may decide to go to the outside for the skills that you need. Once you have identified the needed skills, you will be in a position to select properly from the people available.

To assess the defined needs, you may wish to go back to some of the information and approaches in earlier chapters of this book, and you may wish to use the model of factors affecting safety effectiveness given earlier (p. 192). The various items listed under the three components of behavioral, safety system, and physical environment might each generate a defined need.

ANALYZING A CONSULTANT'S PERFORMANCE

After having worked with a consultant for a period of time, you can judge his or her performance and worth to your organization much more accurately (Figure 14-7).

What are your defined needs?
What skills are needed?
Who has those skills?
 —Externally
 —Internally
What will it cost?

Figure 14-6. Choosing a safety consultant.

Did the consultant:
  Diagnose the problem and define it?
  Consider possible alternatives for solution?
  Suggest the best of these?
  Present his or her proposed solution in a way you could understand?
  Point out the side benefits to the proposed solution?
  Make sure you understood, including what it would cost?

What approach did the consultant use? (See Chapter 11 for explanation.)
  Advocate?
  Expert?
  Alternative identifier?
  Process specialist?
  Reflector?
  Was his or her approach the correct one for your situation?

Answer these questions (see this chapter):
  Did the consultant identify the aim and scope of the study?
  Did he or she submit progress reports?
  Was the consultant professional in approach?
  Did the consultant work well with your people?
  Did your people work well with the consultant?
  Did the consultant teach you and your people?
  Are you/your people more competent as a result?
  Did the consultant achieve his or her objectives?
  Did the proposed program include your/your people's ideas?
  Were you satisfied in how the consultant presented suggestions?
  Was close contact maintained with you throughout?
  Was it accomplished in reasonable time and cost?
  How do your executives rate the consultant?
  Would you retain him or her again?

Did the consultant get results? Explain the indicators.

Figure 14-7. Analyzing your safety consultant.

The following is from an article in *Business Management* on how to conduct a "postmortem" on a management consultant. The article explains:[3]

> Any company that decides to hire a management consultant to solve problems faces problems. Which consultant should it retain? How should it work with him? How can it best implement his recommendations?
>
> Once they've resolved these problems, however, many firms believe their work is done. They overlook or pay only passing attention to one of the biggest problems of all: how to tell whether the consultant really helped the company.

---

[3]Reprinted with permission from How to conduct a post-mortem on a management consultant, *Business Management*, 1965.

Sources of Help

Don't treat this question casually. To conclude an engagement with a consultant with only a vague, general impression that he did a good job or a poor job isn't fair to him or you. . . . Often, a company can't accurately evaluate a consultant's work immediately after he leaves. It may be that an accurate reading can be taken only after his recommendations have been implemented.

Once such an evaluation is undertaken, what questions should be asked? At least these 15:

1. Did the consultant prepare, preferably in writing, some kind of statement outlining the aim and scope of his work, his general plan of procedure, the kind of results he expected to achieve, and the terms of his engagement? Did he then review this report with you in person?

In answering this question, bear this thought in mind: In going over this statement with the consultant, you should have spelled out any limitations you wanted to impose. For example, if your company had certain mediocre employees that you preferred retaining to firing, you should have made this plain. If you didn't indicate any limitations, then you may have been responsible if the consultant made recommendations you didn't want to implement.

So far as the terms of the engagement go, the consultant should have prepared a formal written instrument stating the precise nature of his services, his and your responsibilities, the size of his fee, the method by which he was to be paid, and the duration of the job. This instrument should also have given you the right to terminate the agreement at any time. Such instruments usually take the form of letters of agreement.

If you have not been fully satisfied with the consultants you have employed, your dissatisfaction may be traced to lack of just such a letter. Unwritten agreements breed misunderstanding.

2. Did the consultant carefully plan the work he was to perform?

Ordinarily, a consultant will prepare a plan of study and review it with a client in advance. This study will not only cover the scope of the project, the methods to be employed and the anticipated results, but will also indicate who will do the work, who will supervise this work, how long the job will take, and the type of report that will be submitted.

3. Did the consultant submit progress reports to appropriate executives in your company?

Progress reports should have assured you that the project was proceeding on schedule—or alerted you to deviations or unforeseen trouble. If you ran into the latter, the reports should have enabled you to expand, contract or amend the scope of the consultant's work.

Progress reports should also have enabled you to carry out the consultant's recommendations at once, if that was desirable. And if you went over the reports with the consultant, they should have enabled both of you to crystallize your thinking.

4. Did the consultant's staff operate in a professional manner?

They should have done a workmanlike job of fact finding and analyzing. They should also have demonstrated professional competence, objectivity and integrity.

5. Did the consultant and his staff work constructively with your personnel?

A consulting engagement is always a joint undertaking. The amount of active participation required of a client varies according to the nature of the project. But some participation is almost always necessary.

Ask yourself, therefore, whether the consultant used your employees constructively and, if not, whether he was justified in using only his own staff. Did he also supervise the job adequately? And did he carry it out without unduly disrupting your organization?

6. Were you satisfied that your own organization worked well with the consultant?

If not, make sure you were not to blame. A consulting engagement represents change, disturbance of accepted routine, probably criticism. If you wanted results, you should have paved the way—i.e., set up a single executive to act as a strong liaison with the consultant and indoctrinated your employees about the purpose of the work, its relationship to their jobs, and the importance of their cooperation. If you didn't do this, your employees' morale may have sagged, and they may have resisted the consultant's best efforts to work with them constructively.

7. Did the consultant teach your employees any principles, methods, skills or techniques so that they could manage the improvements he'd suggested once he'd left? Did he stimulate their thinking and expose them to new ideas?

The consultant should have left behind him not only a solution to your immediate problem, but also an accretion in management skills. If he went beyond the problem on which he was asked to work and demonstrated new ways of attacking problems, new questions to ask, new skills and techniques, he left a useful heritage. In the largest sense, the consultant is a teacher, and you should view him in that light.

8. Are you and your associates more competent managers as a result of having worked with the consultant?

It isn't just your middle managers and lower-level employees who should have benefited from the consultant. So, in all likelihood, should you and other members of top management.

9. Did the study achieve your objectives?

The consultant's suggestions should have been practical, timely and suited to your particular needs. They should also have been effective, economical, within the capacity of your employees to carry out, and in accordance with the objectives, policies and long-range plans of your company.

They should not have left you feeling like this dissatisfied client: "All we got was a nice report. The improvements suggested were too theoretical. We were really at a loss when we tried to figure out how to implement them. There was no way to make them pay off in practical results in the shop."

10. Did the recommendations incorporate the best collective judgment of the consultant and your own executives?

The solution to most business problems requires the cooperation of consultant and client. Therefore, the consultant's findings and conclusions should have been developed with your help. They will have practical application only to the extent that your executives reviewed, understood and challenged them—then accepted those that were practical, timely, and suited to the individual requirements of the company.

11. Were you satisfied with the way the consultant's findings and recommendations were reported?

Normally, a consultant indicates in advance what kind of report he will prepare—a full, formal report, a digest report, a report by letter, a report given visually and orally, or some combination of these forms. You, of course, should have approved his intentions. Whatever the nature of his report, he should have taken pains to explain it adequately to those employees responsible for reviewing, accepting and implementing it.

12. Did the consultant maintain close contact with your organization as it considered, then implemented, his recommendations?

This, of course, may not have been necessary. Your employees may have been capable of carrying out his recommendations unaided.

But if that was the case, it may have been desirable for the consultant to make a postinstallation review of the changes or improvements he had recommended. This review should have been made anywhere from three to 12 months after he completed his work.

Such reviews help safeguard any company's investment of time and money. And they allow a consultant to correct misunderstandings, make any necessary modifications and apply recent advances in ideas, techniques or equipment.

13. Was the consultant's work accomplished within a reasonable period of time, for a reasonable fee?

Whether the amount of time was reasonable depends primarily on how well it jibed with the consultant's original estimate. If it didn't jibe, then the extension should have been reasonable and cleared with you in advance.

What's a reasonable fee? Management consultants commonly charge on a per diem basis, often stipulating that their fees will not exceed a certain figure.

In addition, consulting firms normally charge for out-of-pocket expenses—e.g., travel and living expenses, as well as the cost of stenographic, printing and similar services. To use a rough rule of thumb, these expenses should probably have ranged from 10% to 15% of the total of the professional fees.

Again, the over-all cost of the engagement should not have been far out of line with the consultant's original estimate. If it was, he should have discussed this with you in detail before his work was finished—and obtained your approval to proceed.

14. How do your executives as a group evaluate the worth of the consultant's work?

Some obvious points to look into: How many of the consultant's recommendations were accepted? How many were rejected? How many were modified? What was the operating impact of the engagement on your organization?

15. Would you retain the consultant again if you had need of his kind of services?

This, obviously, is the key question. All the others lead up to it.

It's possible that your consultant may have measured up to most of the above criteria, yet left you feeling you wouldn't want to use him again. If that's the case and if you have conducted your post-mortem objectively, then you have indeed conducted a post-mortem for that consultant.

It may be that as a result of your analysis of what you are getting, you will decide that perhaps you can do as well internally. Many companies use the internal approach now, and more are turning to it, in both safety and nonsafety areas.

INTERNAL PROBLEM SOLVING

K. Albert, in his book *How to Be Your Own Management Consultant,* suggests that there are four different types of internal management problem-solving approaches:

1. To hire a full-time internal consultant.
2. To put someone on a special assignment on a temporary basis.
3. To create a task force to work on a problem.
4. Collaboration.

Each is somewhat self-explanatory except the last. Collaboration is a partnership between an outside consultant and an internal consultant. Albert suggests—no matter which approach is chosen—these ground rules for success:

1. Get total support of top management.
2. Earn the acceptance of operation units.
3. Report in to a high level.
4. Establish confidentiality.
5. Avoid company politics.
6. Maintain objectivity.
7. Start slowly.

PART IV | Appendixes

Appendix A

Measurement

To hold someone accountable, we must know whether he or she is performing well; so we must measure that person's performance. Without measurement, accountability becomes an empty and meaningless concept. Then, because we propose that the single most important factor in getting good line safety performance is accountability, we find that we often are really talking about ways to measure the line manager better. And measurement has been our downfall in safety for years.

For the line manager, "to measure is to motivate." Although this statement might have sounded a little ridiculous 20 years ago, I believe that it expresses a profound truth, at least in terms of the safety performance of line organizations. Managers react to the measures used by the boss; they perceive a task to be important only when the boss thinks it is worth measuring.

Having perceived the importance of measurement in obtaining good safety performance, we then hit our biggest snag: What will we measure? Should we measure our failures as demonstrated by accidents that have occurred in the past? If this is in fact a good measure, as has historically been believed (for that is what we usually measure), then what level of failure should be measured? We can measure the level of failure we call "fatalities." Fatalities are used to measure our national highway traffic safety endeavors. Is the measure of fatalities, then, a "good" measure? Fatalities might be a good measure if we were assessing the national traffic safety picture, but would be a little ridiculous in the case of a supervisor of ten factory workers. Such a supervisor might well do absolutely nothing to promote safety and still never experience a fatality in his or her department. Obviously, measuring fatalities would make little sense in this case.

Although this may sound a little ridiculous, it accurately describes what is going on in many safety programs today. A supervisor can do absolutely nothing related to safety for a year and attain a zero frequency rate with a small bit of luck. By rewarding such a supervisor, we are actually reinforcing nonperformance in safety.

If fatalities (or frequency rates) are a poor measure of supervisory performance, what is a good measure? Or, more important, what is wrong with fatalities as a measure? Perhaps measuring our failures is not the best approach to use in judging safety performance. After all, this is not the way we measure people in other aspects of their jobs. We do not, for instance, measure line managers by the number of parts the people in their departments failed to make yesterday. And we do not measure the worth of salespeople by the number of sales they did not make. Rather, in cases like these we decide what performances we want and then we measure to see whether we are getting them.

What would be a good measure of supervisory safety performance? More important, what set of criteria can we develop for measuring supervisory safety performance? Or the safety performance of the corporation? Or our national traffic safety performance? Or anything else related to safety?

Even a brief look at the problem of measurement shows us that we need different measures for different levels in an organization, for different functions, and perhaps even for different managers. What is a good measure for one supervisor of ten people may not be a good measure for another, much less for a plant superintendent or the general manager of seven plants and 10,000 people. What might be a good measure for the supervisor of a foundry cleaning room may be inappropriate for use in judging the effectiveness of OSHA.

Available Measures

There are two categories of available tools: (1) activity (or performance) measures and (2) results measures. (See Figure A-1.) As a general rule, in selecting measuring devices, use only activity measures at the lower managerial levels and primarily activity measures (with some results measures) at the middle–upper management levels, saving the pure results measures for the executive level. With rare exceptions (such as safety sampling), this rule of thumb gives us statistical validity and is extremely important.

In choosing either activity measures or results measures to determine performance, we can use them at the supervisory, managerial, or systemwide levels, provided we use caution in measurement selection. (See Figure A-2.) At the supervisory level, the activity measures are most appropriate, whether it be the number of inspections, number of people trained, or number of observations made. These activity measures are equally appropriate at managerial levels, and can even be used at systemwide levels (through audits or questionnaires). All activity measures are extremely well suited to MBO (SBO) approaches (were the objectives

Measurement

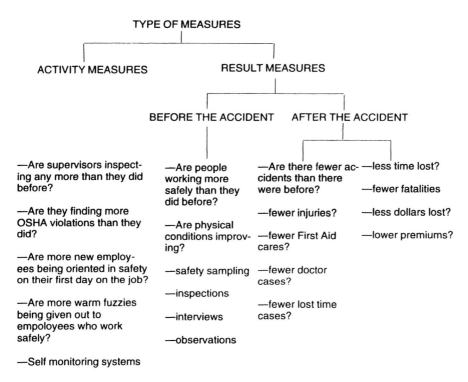

Figure A-1. Available measures.

reached?). Using these measures in SBO assumes that the original objectives were well written (realistic, rifled, measurable, and under the direct control of the objective setter).

Results measures also can be used at all levels, as long as extreme care is used at the lower levels. The traditional safety measures such as frequency rate or severity rate cannot be used at the lower levels except over long periods of time and then probably only as a quality check.

Measures at the Supervisory Level

Performance Measures

We start our examination of supervisory measurement by looking at what the supervisor does to get results and by determining whether the supervisor actually does them.

Performance measures have certain distinct advantages:

ACTIVITY		
SUPERVISOR For: Objectives Met	MANAGER Objectives Met	SYSTEM-WIDE Audit
# Inspections # Quality Investigations # Trained # Hazard Hunts # Observations # Quality Circles	Use of Media # Job Safety Analyses # Job Safety Observations # One-on-One # Positive Reinforcement Group Involvement	—Questionnaires —Interviews
RESULTS		

FOR:
SUPERVISORS
Inspection Results
	MANAGERS
	Safety Sampling
	Inspection Results
	Safety Performance
	  Indicator
	Estimated Costs
	Control Charts
	Property Damage
		SYSTEMWIDE
		Safety Sampling
		Safety Performance
		  Indicator
		# First Aid or Frequency
		# Near Misses or Frequency
		Property Damage
		Frequency–Severity Index
		Estimated Cost Control
		  Charts

Figure A-2. Activities and results measures for supervisors, managers, and systemwide safety programs.

- They are flexible and take into account individual supervisory styles. We do not have to use the same measure for all supervisors. We can allow each supervisor to select performances and levels of performance and then measure these performances and levels. Performance measures are excellent for use in MBO approaches.
- They give swift feedback, as most require supervisors to report their level of performance to the boss. (They are also often self-monitoring.)
- They all measure the presence, rather than the absence, of safety.
- They are usually simple and thus administratively feasible.
- They are clearly the most valid of all measures.

Failure Measures

Failure measures tend to be generated from our injury record system. In order for injury records to be used for measuring the results of the line manager's safety performance, they should be set up so that:

- They are broken down by unit.
- They give some insight into the nature and causes of accidents.

- They are expressed eventually in terms of dollars by unit.
- They conform to any legal and insurance requirements.

Beyond these broad outlines, each company can devise any system that seems right for it. The "dollar" criterion is included because of the belief that a dollar measuring stick is much more meaningful to line personnel than any safety specialist's measuring stick (such as our frequency and severity rates).

Before-the-Fact Measures

Before-the-fact measures measure the results of supervisor action before an accident occurs. For instance, a periodic inspection is made of a supervisor's work area to measure how well he or she is maintaining physical conditions. This is a measure of whether things are wrong and, if so, how many things are wrong. We can also measure how well a supervisor gets through to the people in the department by measuring the people's work behavior (safety sampling).

One of the more recent and one of the best measurement tools is the perception survey.

Measures at Middle- and Upper- Management Levels

Performance Measures

Although we begin to emphasize results measures more at the middle-management level, some performance measures should be retained here also. Performance measures at this level are simple measures of what middle managers do—of whether they perform the tasks that it is necessary for them to perform. Thus, everything we have said about supervisory performance measures applies also at the middle-management level.

However, supervisory performance and middle-management performance should be quite different, necessitating some changes in the measures used at this level. Normally, we want the middle managers to get their subordinates (supervisors) to do something in safety. Thus, we can measure them to see whether they meet with their supervisors, check on these supervisors, or monitor the quality of the supervisors' work.

Audits

A measure of performance (and results) that enters the scene at the middle- and upper-management levels (management of a location, for instance) is the audit of safety performance.

ACCOUNTABILITY FOR ACTIVITIES	
MANAGEMENT MEASURES WHAT SUPERVISOR IS DOING	
(1) Safety meetings held	(5) Accident investigation
(2) Tool box meetings	(6) Incident reports
(3) Activity reports on safety	(7) Job hazard analyses
(4) Inspection results	(8) Any defined tool
MANAGEMENT MEASURING TOOLS	
(1) Safety sampling	(6) SCRAPE
(2) Statistical controls	(7) Performance ratings
(3) Critical incident techniques	(8) Objectives met (SBO)
(4) Safe-T-scores	(9) Audits
(5) Regular reports	
ACCOUNTABILITY FOR RESULTS	
(1) Charge accidents to departments	(5) Number of incidents
(a) Charge claim costs to the line	(a) Accidents
(b) Include accident costs in the profit and loss statements.	(b) Injuries
	(c) Other
(2) Prorate insurance premiums	(6) Costs
(3) Put safety into the supervisor's appraisal	(7) Frequency and severity indicators
(4) Have safety affect the supervisor's income	(8) Estimate costs
	(9) Loss ratios
	(10) Costs of damage

Figure A-3. Accountability for activities and for results.

Results Measures at Middle and Upper Management

At the middle-management level we concentrate on results measures. Here also we can work with before-the-fact measures or with failure measures. The validity and reliability of our failure measures may depend upon the size of the unit we are working with. Before-the-fact results measures used to judge a middle manager's performance would include safety sampling and inspections by staff safety. Failure measures can be used a bit more extensively at this level. Here our traditional indicators become more useful. (See Figure A-3.)

Appendix B

Safety Sampling

One of the newer methods of fixing accountability, using statistical methods, is safety sampling. Safety sampling measures the effectiveness of the line manager's safety activities, but not in terms of accidents. It measures effectiveness before the accident by taking a periodic reading of how safely the employees are working.

Like all good accountability systems or measurement tools, safety sampling is also an excellent motivational tool, for each line supervisor wants to be sure the employees are working as safely as possible when the sample is taken. To accomplish this, he or she must carry out some safety activities such as training, supervising, inspecting, and disciplining.

In those organizations that have utilized safety sampling, many report a good improvement in their safety record as a result of the improved interest in safety on the part of the line supervisors.

Safety sampling is based on the quality control principle of random sampling inspection, which is widely used by inspection departments to determine quality of production output without making 100 percent inspections. For many years, industry has used this inspection technique, in which a random sampling of a number of objects is carefully inspected to determine the probable quality of the entire production. The degree of accuracy desired dictates the number of random items selected, which must be carefully inspected. The greater the number inspected, the greater the accuracy.

Procedure

I. Prepare a Code

The list of unsafe practices is the key to safety sampling and supervisor training. This list contains specific unsafe acts that occur in the plant. These are the "accidents about to happen." The list is developed from the accident record of each plant. In addition, possible causes are also listed. The code is then placed on an observation form (see Figure B-1).

II. Take the Sample

With the code attached to a clipboard and with a theater counter, the sampling begins. The inspector identifies the department and the supervisor responsible. He or she then proceeds through the area, observing every employee who is engaged in some form of activity. The inspector instantaneously records a safe or an unsafe observation for the employee.

Each employee is observed only long enough to make a determination, and once the observation is recorded, it should not be changed. If the observation of the employee indicates safe performance of the job, it is counted on the theater counter. If the employee is observed performing an unsafe practice, a check is made in the column that indicates the type of unsafe practice by the element code number (see Figure B-1).

III. Validate the Sample

The number of observations required to validate is based on a preliminary survey and the degree of desired accuracy. The following data must be recorded on the preliminary survey: total observations and unsafe observations. The percentage of unsafe observations is then calculated. Using this percentage (P) and the desired accuracy, which we will determine as $Y \pm 10\%$, we can calculate the number of observations (N) required by using the following formula:

$$N = \frac{4(1-P)}{Y^2(P)}$$

where N = total number of observations required
P = percentage of unsafe observations
Y = desired accuracy

For example, if the preliminary survey produced the following results: (1) total observations 126 and (2) unsafe operations 32, the percentage of unsafe to total

Safety Sampling

SAMPLING WORKSHEET
Page 1 of 1

Safe observations
Unsafe acts

DEPARTMENT

#	Unsafe act	DC & Service	Maint. Power	Tool room	Foundry & pattern	Stock & Shipping	Rotor	Shaft	Punch Press	Body & frame	Bracket	Small winding	Large winding	Small assembly	Lg. assem. & pck.
1	Improper lifting														
2	Carrying heavy load														
3	Incorrect gripping														
4	Lifting without protective wear														
5	Reaching to lift														
6	Lifting and turning														
7	Lifting and bending														
8	Improper grinding														
9	Improper pouring														
10	Swinging tool toward body														
11	Improper eye protection														
12	Improper foot wear														
13	Loose clothing—moving parts														
14	No hair net or cap														
15	Wearing rings														
16	Finger/hands under dies														
17	Operating equip. at unsafe speeds														
18	Foot pedal unguarded														
19	Failure to use guard														
20	Guard adjusted improperly														
21	Climbing on machines														
22	Reaching into machines														
23	Standing in front of machine														
24	Leaning on running machine														
25	Not using push sticks (jigs)														
26	Failure to use hand tools														
27	Walking under load														
28	Leaning—suspended load														
29	Improper use of compressed air														
30	Carrying by lead wires														
31	Table too crowded														

Figure B-1. Sampling worksheet.

	DEPARTMENT													
SAMPLING WORKSHEET Page 1 of 1 Safe observations	DC & Service	Maint. Power	Tool room	Foundry & pattern	Stock & Shipping	Rotor	Shaft	Punch Press	Body & frame	Bracket	Small winding	Large winding	Small assembly	Lg. assem. & pck.
Unsafe acts														
32 Hands and fingers between metal boxes														
33 Underground power tools														
34 Grinding on tool rest														
35 Careless alum. splash														
36 One bracket in shaft piling														
37 Feet under carts or loads														
38 Pushing carts improperly														
39 Pulling carts improperly														
40 Hands or feet outside lift truck														
41 Loose material under foot														
42 Improper piling of material														
43 Unsafe loading of trucks														
44 Unsafe loading of skids														
45 Unsafe loading of racks														
46 Unsafe loading of conveyers														
47 Using defective equipment														
48 Using defective tools														
49 Evidence of horseplay														
50 Running in area														
51 Repair moving machines														
52 No lock-out on machine														
TOTAL UNSAFE ACTS														
ADDITIONAL UNSAFE ACTS:														
53														
54														
55														
56														
57														
58														

Date _____ Time: _____ Sampler: _____

Figure B-1. *(cont.)*

Safety Sampling

observations would be 32 divided by 126, which is 0.254, times 100, or 25%. Thus:

$$N = \frac{4(1-P)}{Y^2(P)}$$

$$N = \frac{4(1-0.25)}{(0.10)^2(0.25)}$$

$$N = \frac{3}{0.0025}$$

$N = 1{,}200$ (no. of observations required)

To give effective results, this study must have a minimum of 1,200 observations. (See Figure B-2.)

Percentage of unsafe observations	Observations needed	Percentage of unsafe observations	Observations needed
10	3,600	30	935
11	3,240	31	890
12	2,930	32	850
13	2,680	33	810
14	2,460	34	775
15	2,270	35	745
16	2,100	36	710
17	1,950	37	680
18	1,830	38	655
19	1,710	39	625
20	1,600	40	600
21	1,510	41	575
22	1,420	42	550
23	1,340	43	530
24	1,270	44	510
25	1,200	45	490
26	1,140	46	470
27	1,080	47	450
28	1,030	48	425
29	980	50	400

Figure B-2. Observation needed vs. percentage of unsafe observations.

IV. Report to Management

The results can be presented in many different forms (see, e.g., Figures B-3 through B-6). However, the report should include the following:

1. Total percentage of unsafe activity by department and by shift.
2. Percentage of unsafe activity by supervisor, general foreman, or superintendent.
3. Number and type of unsafe practices observed.
4. Types and number of unsafe observations that are supervisory responsibilities.

SAFE PRACTICE SAMPLING REPORT

Plant ___1___   Period covered ___October___

Department Supervisor	Unsafe practice code number												Observations		Percentage unsafe
	1	2	7	9	11	17	26	34	36	59			Total	Unsafe	
E. Jones–supt.													1,094	39	3.4
Smith–gen. for.													246	9	3.5
Jolas				1			1	1					90	3	3.2
Johnson		3			1		1	1					156	6	3.8
G. McArthur													226	11	4.6
Mantle			1		1		1	1					101	4	3.8
Williams			1	1									53	2	3.6
Nedstrom				1			1	1	2				72	5	6.5
Mack													284	13	4.4
Peters		1			3	1							96	5	5.0
Sadeiri						3	1	1					73	5	6.4
Altert	1		1										64	2	3.0
Anderson	1												51	1	1.9

Figure B-3. Safe practice sampling report.

SAFETY SAMPLING REPORT

Plant ___1___ Month of ___October___

Department	Total observations	Unsafe observations	Percentage of unsafe activity	
			This month	Previous month
Manufacturing-Prod	442	77	17.4	12.1
Press	1,815	244	15.3	19.7
Assembly	1,699	59	4.0	4.0
Welding	322	70	21.0	11.2
Subtotals	4,278	450	14.4	14.2
Production Eng.	339	55	16.2	21.5
Plant Engineering	341	51	14.9	26.7
Subtotals	680	106	15.6	23.6
Plant totals				

Figure B-4. Safety sampling report.

SUMMARY OF SAMPLING

Period covered ___October___

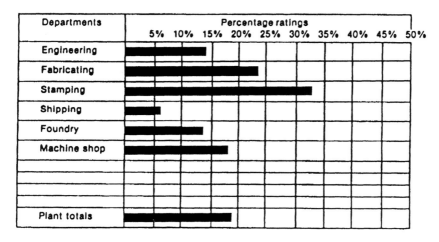

Figure B-5. Summary of sampling.

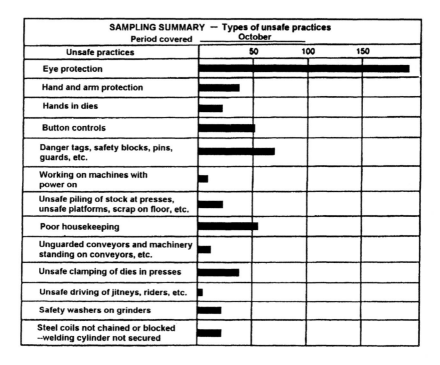

Figure B-6. Sampling summary—types of unsafe practices.

Appendix C

Accountability Systems

Any accountability system that defines, validly measures, and adequately rewards will work. This appendix gives three examples.

SCRAPE

SCRAPE is an acronym for a Systematic method of Counting and Rating the Accident Prevention Effort. Most companies measure accountability through analysis of results; monthly accident reports at most plants suggest that supervisors are judged by the number and the cost of accidents that occur under their jurisdiction. We should judge line supervisors by what they do to control losses, and SCRAPE is a simple way of doing this. It is as simple as deciding what supervisors are to do and then measuring to see that they do it.

The SCRAPE rate indicates the amount of work done by a supervisor or a manager to prevent accidents in a given period. Its purpose is to provide a tool for management that shows, before the accident, whether or not positive means are being used regularly to control losses.

The first step in SCRAPE is to determine specifically what the line managers are to do in safety. Normally this falls into the categories of (1) making physical inspections of the department, (2) training or coaching people, (3) investigating accidents, (4) attending meetings of the workers' boss, (5) establishing safety contacts with people, and (6) orienting new people.

With SCRAPE, management selects which of these things it wants supervisors to do and then determines their relative importance by assigning values to each.

Item	Points
Departmental inspections	25
Training or coaching (e.g., 5-minute safety talks)	25
Accident investigations	20
Individual contacts	20
Meetings	5
Orientation	5
Total	100

Figure C-1. Point values of items, SCRAPE.

Let us suppose that management believes the six items above are the things it wants supervisors to do and believes that investigations and individual employee contacts, etc., are most important. Attending meetings and orienting new people are relatively less important at this time. Management might then assign the values shown in Figure C-1. Depending on management's desires, the point values can be increased or decreased for each item.

Every week each supervisor will fill out a small form (Figure C-2) indicating activity for the week. Management, on the basis of this form, spotchecks the quality of the work done in all size areas, and rates the accident prevention effort by assigning points between zero and the maximum. For example:

- In Department A, the supervisor makes an inspection and makes six corrections. The boss later inspects, finding good physical conditions. Supervisor A rates the maximum of 25 points.

Department _____	Week of _____	Points
1. Inspection made on _____	# corrections _____	_____
2. 5-minute safety talk on _____	# present _____	
3. # accidents _____	# investigated _____	
Corrections _____		_____
4. Individual contacts: Names _____		
		_____
5. Management meeting attended on _____		_____
6. New people (names): _____	Oriented on (dates): _____	
_____		_____

Figure C-2. Weekly activity form, SCRAPE.

Accountability Systems

- In Department B, the supervisor makes an inspection but no corrections. The plantwide inspection, however, indicates that much improvement is needed. Supervisor B might get only 5 points for making the inspection but doing a poor job of it.
- In Department C, there were five accidents. Only one individual lost time, and Supervisor C turned in only one investigation, getting only 5 points for the effort.
- In Department D, there are 43 employees, but only 3 were individually contacted during the week. This might also be worth only 5 points.

Management decides relative values by setting maximum points and also sets the ground rules about how maximum points can be obtained. Each week a report is issued (Figure C-3).

SCRAPE can provide management with information on how the company is performing in accident prevention. It measures safety activity, not a lack of safety. It measures before the accidents, not after. Most important, it makes management define what it wants from supervisors in safety and then measures to see that it is achieving what it wants. SCRAPE is a system of accountability for activities.

SAFETY BY OBJECTIVES

Historically, our safety programs have failed in some essential elements. Many programs are far from producing behavior that could be considered goal-directed. Responsibilities, even with written policy, are often unclear. Participa-

Week of								
	ACTIVITY							
Department	Inspect (25)	5-minute talks (25)	Accident invest. (20)	Indiv. contacts (20)	Meet atten (5)	Orient (5)	Total rate (100)
A	25	15	20	15	5	5	85
B	5	10	20	5	5	5	50
C	25	10	5	5	5	5	55
D	15	25	20	20	—	5	85
E	10	5	—	—	5	—	20
F	20	20	15	5	—	—	60
Average	17	14	13	8	3	3	58

Figure C-3. Weekly report, SCRAPE.

Organization	Size	Type	Results
Food Processor	15,000 employees	Delimited, with both mandatory & optional	36% and 40% red. yrs. 1 & 2
Food Processor	10,000 employees	Delimited, with both mandatory & optional	30% and 25% red. yrs. 1 & 2
Contractor	3,000 employees		67% red. 2 yrs
Brewery	15,000 employees	1/2 limited 1/2 no limits	64% red. 3 years
Oil Producrion	3,000 employees	Limited	50% red. 1st year
Railroad	4,000 employees	Limited	50% red. 1st year
Metal Manufacturer	3,000 employees	Select from a list of 12	50% red. 1st year

Figure C-4. Accident frequency results for several organizations using SBO approach.

tion in goal setting and decision making is almost nonexistent. Feedback and reinforcement are slow and often not connected to the amount of effort expended in safety (especially when the number or severity of accidents is the measuring stick). Planning is minimal, and although results often are measured, freedom of decision or control seldom is left to the lower levels. Finally, imagination and creativity are rare commodities in most safety programs. The principles of MBO (management by objectives), adapted to safety programming (SBO), can overcome some of these failings.

These are the steps of SBO:

1. Obtain management–supervision agreement on objectives. In an SBO program, the agreement should emphasize only activities objectives. Initially, then, the agreement reached will be on strategies and objectives (what means, tools, and resources are to be used).
2. Give each supervisor an opportunity to perform. Once agreements are completed, leave supervisors alone to proceed with their action plans. Require only progress reports.
3. Let supervisors know how they are doing. With quantified objectives (activity objectives must be quantified) give regular, current, and pertinent feedback so they can adjust their plans when they see the need.
4. Help, guide, and train. Both management and safety staff fulfill this role. Safety staff provides the technical and safety technical expertise. Management provides the managerial help when asked, guidance, and training at the outset.
5. Reward supervisors according to their progress. This requires a reward system that is geared to the progress made toward agreed-upon objectives. Various managerial rewards should be used: pay, status, advancement, recognitions, etc.

The SBO approach has been implemented in a variety of settings: brewing (Adolph Coors Company), chemical (DuPont), railroading (Frisco, Union Pacific,

Accountability Systems   *251*

Santa Fe, Chicago Northwestern), paper (Hoerner-Waldorf), and other industries. The results, of course, are not uniform. Some are successful; some are not. Much depends on the installations, the commitment of management, and the meaningfulness of the objectives that were set.

See Figure C-4 for results in some of the organizations using SBO. It has, of course, been installed differently in each organization. Some have left the objective-limited supervisory choices to certain allowable strategies. One organization's management spelled out 12 areas that were believed to be areas of slippage in the safety program.

For the most part, SBO works and continues to work once implemented. One of the above companies reported a 75 percent reduction in the frequency rate in the first three months of the program, one reported a six-month claim savings of $2.3 million, and one reported a 67 percent reduction in frequency continuing after three years.

MENU

A MENU system defines certain tasks as mandatory, as in SCRAPE. In addition, supervisors can use a number of other tasks if they choose to. From the menu that is provided, they must select some to use. In the optional tasks, they also must set numerical activity goals that they can be measured against.

Figure C-5 through C-7 show examples of the definition of accountabilities for the first line supervisors in two different organizations and a weekly safety report, all in a Menu system.

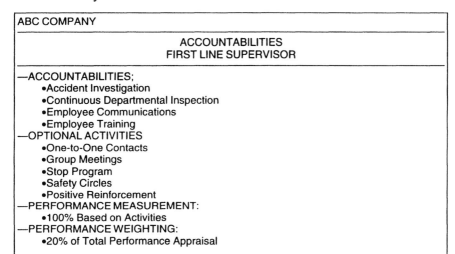

Figure C-5.   Accountabilities of first line supervisor in ABC Company.

XYZ COMPANY	
	ACCOUNTABILITIES **FIRST LINE SUPERVISOR**
GENERAL	—The key accountability is to carry out the tasks defined below.
TASKS	—Hold a periodic meeting with all employees. —Include safety status in all work group meetings. —Inspect department weekly and write safety work orders as required. —Have at least five one-to-one contacts regarding safety with employees each week. —Investigate injuries and accidents in accordance with Managing Safety guidelines within 24 hours. In addition, in agreement with Department head: —Select at least two other tasks from a provided list (See Section 4, Subject 4.8) and agree on what measurable performance is acceptable. —Report on these activities weekly.
WEEKLY SAFETY REPORT	—The first line supervisor shall prepare and distribute a Weekly Safety Report in accordance with the format shown in Exhibit 3-5.
MEASURE OF PERFORMANCE	—Successful completion of tasks.
REWARD OF PERFORMANCE	—Safety will be listed as one of the key measures on the Accountability Appraisal Form.

Figure C-6. Accountabilities of first line supervisor in XYZ Company.

FIRST LEVEL MANAGER'S WEEKLY SAFETY REPORT

FROM: _____   TO: _____   WEEK ENDING: _____

1. WORK GROUP SAFETY MEETING
   DATE: SUBJECT:
2. DEPARTMENT SAFETY INSPECTION
   DATE: FINDINGS:
3. ONE-TO-ONE CONTACTS
   EMPLOYEE: DATE:
   EMPLOYEE: DATE:
   EMPLOYEE: DATE:
   EMPLOYEE: DATE:
   EMPLOYEE: DATE:
4. INJURY STATUS
   NAME: DATE:
   INJURY DESCRIPTION:

5. OTHER SAFETY TASKS
   DESCRIPTION   ACTION   % COMPLIANCE TO GOAL

REPORT DISTRIBUTION: STAFF II MANAGER
                     EMPLOYEE RELATIONS MANAGER

Figure C-7. A first level manager's weekly safety report.

Appendix *D*

Supervisory Tasks

The required tasks for supervisory performance may include almost anything, from the traditional tasks of the past, to the new and different ones for today, and others sure to emerge in the future. This appendix presents some ideas on traditional and nontraditional supervisory tasks.

THE TRADITIONAL SUPERVISORY TASKS

Investigating

An accident that causes death or serious injury obviously should be thoroughly investigated. A "near-accident" that might have caused death or serious injury is equally important from the safety standpoint, and it too should be investigated (e.g., the breaking of a crane hook or a scaffold rope, or an explosion associated with a pressure vessel).

Any epidemic of minor injuries demands study. A particle of emery in the eye or a scratch from handling sheet metal may be a very simple case. The immediate cause may be obvious, and the loss of time may not exceed a few minutes. However, if cases of this or any other type occur frequently in the plant or in your department, an investigation might be made to determine the underlying causes.

With any accident, we must find the fundamental root causes and remove them if we hope to prevent a recurrence. Root causes often relate to the management system, and may be due to management's policies and procedures, supervision and

```
Name of injured _____ Date of Accident _____ Time _____
Seriousness:  Lost Time   Doctor   First Aid Only   Near Miss
Nature of Injury _____
What Happened? _____
_____

What acts and conditions* were involved? What caused them? How were they corrected?
  *At least five (use back also)
```

Unsafe Act/Condition/ Symptom	Possible/Probable Cause	Correction/ Suggested Correction
1.		
2.		
3.		
4.		
5.		

Supervisor _____ Department _____

Figure D-1. Supervisor's report of accident investigation.

its effectiveness, or training. Root causes are those whose correction would effect permanent results.

Traditionally investigations have asked for the identification of unsafe act and unsafe conditions. They should go much further than this. Figures D-1 and D-2 show accident investigation forms that require an analysis of unsafe acts and conditions and a listing of causes and corrections by the supervisor, with further evaluation by the general foreman.

Inspecting for Hazards

Inspection is one of the primary tools of safety. At one time, it was virtually the only tool, and it still is the one most used. Figure D-3 shows an inspection form.

Job Safety Analysis

Job Safety Analysis (JSA) is a procedure that identifies the hazards associated with each step of a job and develops solutions for each hazard that either eliminate it or control it. Figure D-4 is an example of a job analysis worksheet.

Supervisory Tasks

		Circle One	
1. Was it on time?	Yes - 5 pts.		No - 0 pts.
2. Was seriousness indicated?	Yes - 5 pts.		No - 0 pts.
3. Does it say where it happened?	Yes - 5 pts.		No - 0 pts.
4. Can you tell exactly what the injury is?	Yes - 5 pts.		No - 0 pts.

			Circle One			
5. How many acts and conditions are listed?	5	4	3	2	1	0
6. How many causes are identified?	5	4	3	2	1	0
7. How many corrections were made or suggested?	5	4	3	2	1	0
8. How many of the listed corrections would have prevented this accident?	5	4	3	2	1	0
9. How many corrections are permanent in nature?	5	4	3	2	1	0
10. In how many of the corrections listed is the supervisor now doing something differently?	5	4	3	2	1	0

Total of Circled Points
Multiply × 2
Reviewed by _____   SCORE _____
                General Foreman

Figure D-2. Foreman's evaluation of supervisor's accident investigation report.

Figure D-2. Foreman's evaluation of supervisor's accident investigation report.

Job Safety Observations

A Job Safety Observation (JSO) provides a way to learn more about the work habits of people. Following the procedure shown in Figure D-5, you can use this opportunity to check on the results of past training; make immediate, on-the-spot corrections and improvements in work practices; and compliment and reinforce

Supervisor's Inspection Form		
Name _____		Date _____
Symptom Noted Act/Condition/Problem	Causes Why—What's Wrong	Corrections Made or Suggested By you—By others

Figure D-3. Supervisor's inspection form.

Job: Using a Pressurized Water Fire Extinguisher

WHAT TO DO (Steps in sequence)	HOW TO DO IT (Instructions) (Reverse hands for left-handed operator.)	KEY POINTS (Items to be emphasized. Safety is always a key point.)
1. Remove extinguisher from wall bracket.	1. Left hand on bottom lip, fingers curled around lip, palm up. Right hand on carrying handle palm down, fingers around carrying handle only.	1. Check air pressure to make certain extinguisher is charged. Stand close to extingusiher, pull straight out. Have firm grip, to prevent dropping on feet. Lower, and as you do remove left hand from lip.
2. Carry to fire.	2. Carry in right hand, upright position.	2. Extinguisher should hang down alongside leg. (This makes it easy to carry and reduces possibility of strain.
3. Remove pin.	3. Set extinguisher down in upright position. Place left hand on top of extinguisher, pull out pin with right hand.	3. Hold extinguisher steady with left hand. Do not exert pressure on discharge lever as you remove pin.
4. Squeeze discharge lever.	4. Place right hand over carrying handle with fingers curled around operating lever handle while grasping discharge hose near nozzle with left hand.	4. Have firm grip on handle to steady extinguisher.
5. Apply water stream to fire.	5. Direct water stream at base of fire.	5. Work from side to side or around fire. After extinguishing flames, spray water on smouldering or glowing surfaces.
6. Return extinguisher. Report use.		

Figure D-4. Job analysis worksheet.

JOB SAFETY OBSERVATION		
Employee:	Supervisor:	
Job:	Date:	Time:
Notes on Any Job Practices that Are Unsafe:		
Notes on Any Practices that Need Change or Improvement:		
Notes on Any Practices that Deserve Complimenting:		
Notes on the Review and Discussion:		

Figure D-5. Worksheet for job safety observation.

safe behavior. Through your comments you can encourage proper attitudes toward safety.

One-on-Ones

While slightly less traditional than some of the procedures mentioned, one-on-ones are becoming more common and popular every year. There is a good rationale behind the one-on-one contact. It makes much more sense than the traditional safety meeting; it allows leveling, it allows the supervisor to understand the worker's needs (which is essential to motivation), and it provides a way to break through the power of the informal group.

Most organizations utilizing one-on-ones set up some structure to ensure that they do in fact happen. This may be a quota system or some other mechanism or report.

THE NONTRADITIONAL SUPERVISORY TASKS

Many nontraditional tools are being used today. We mention only a few here to give an idea of the possibilities.

Positive Reinforcement

One of the best approaches is to force the utilization of positive reinforcement. Following the contacts and observations mentioned above, many organizations are requiring supervisors to provide some positive strokes when safe behavior has been observed. There have been many experiments on the use of positive reinforcement, and all to date have shown excellent results. A recent study of some 35 of these experiments showed an average improvement in safe behavior of 60 percent in a relatively short time period. This therefore appears to be a powerful tool for improving safe behavior.

Safety Sampling

Safety Sampling (SS) is a well-tested technique in accident prevention. It is a little different from the other techniques described in this section in that it normally is implemented on a companywide basis by management or the safety director rather than within one department by the supervisor. Nevertheless, because it has been so effective in safety programs, we believe a supervisor can use this tool advantageously.

Incident Recall Technique (IRT)

Confidential attitude surveys conducted by consulting firms in a number of companies have revealed that it is fairly common for supervisors to hide accidents. It is even more common for employees to hide them. Therefore, the basic objective of the IRT is to gain the cooperation of the employee, so that he or she can and will freely relate all incidents from the past that can be recalled. The success of the IRT, in terms of number of incidents revealed, depends primarily on the skill of the interviewer. (See Figure D-6.)

Technique of Operations Review (TOR)

The Technique of Operations Review (TOR) was devised to help find some of the multiple, interrelated causes behind accidents requiring investigation. It is basically a tracing system, but it also can be used as a training technique in safety. The TOR begins with an incident; and its purpose is to expose the real

INCIDENT RECALL TECHNIQUE			
Person Interviewed:			Date:
Supervisor:		Department:	
Incidents Recalled:			
Analysis of Causes of Recalled Incidents:			
Action Taken on Causes:			

Figure D-6. Worksheet for incident recall.

problems behind the incident, which are viewed as symptoms of more serious trouble.

Hazard Hunt

Another method that had been used successfully to spot possible accident causes is the hazard hunt. It is also good for involving people. To implement the procedure follow these steps:

1. Make copies of the hazard hunt form (Figure D-7).
2. Hold a short session with employees to explain the form and the reasons for using it.
3. Have employees jot down anything they feel is a hazard and return the form.
4. Review the forms, correct those hazards that can be corrected, and initiate action on the others.
5. Always inform employees of the actions, even if the hazards they mention are not really problems. Then hash over any disagreements with individuals to clear the air.
6. If something is a hazard that must be corrected, assign a priority to it and schedule it for rectification.

```
To: _____
From: _____
                    HAZARD HUNT
I think the following is a hazard:_____
_____
_____

DO NOT WRITE BELOW HERE—TO BE FILLED IN BY SUPERVISOR
Supervisor:
  Agree this is a hazard.
    Corrected by Supervisor on _____   Discussed on _____
    If cannot correct, sent to Personnel on _____
    Job order on _____  Scheduled _____   Discussed on _____

  Do not agree this is a hazard.
    Discussed _____        Conclusion_____
    To Personnel _____        Conclusion_____
DO NOT WRITE HERE—FOR PERSONNEL USE
Supervisor_____   HH No._____
Matrix No. _____ (Seriousness)
```

Figure D-7. Hazard hunt.

Some Behavioral Tasks

Most of the above supervisory tools have been used in one or more organizations. Other strategies also could be used for getting information and for better handling the worker, such as the three that follow.

Climate Analysis

A quick, superficial climate reading of an organization can be obtained by filling out the climate analysis form (Figure D-8).

Inverse Performance Standards

The traditional way to make performance appraisals is to have the boss evaluate the subordinate. When inverse performance standards are used, the subordinate evaluates the boss. Subordinates fill out a form (see Figure D-9) and submit it

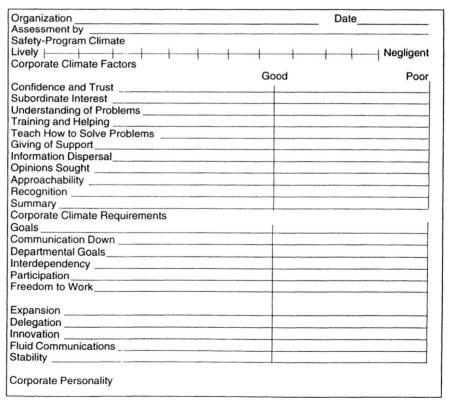

Figure D-8. Climate analysis form.

Supervisory Tasks

anonymously to a third person, who keeps the responses confidential and provides the boss with a summary of all responses.

Worker Safety Analysis

Safety professionals and supervisors are generally aware of the concept (mentioned above) of Job Safety Analysis (JSA), the systematic analysis of specific jobs in order to spot situations with accident potential. Similarly, worker safety analysis is the systematic analysis of a worker.

Management can devise a form (such as Figure D-10) to assist the supervisor in observing each worker to determine whether he or she is highly likely to make human errors. Are there logical reasons why any particular employee is likely to make such errors? Worker safety analysis can uncover these reasons.

```
Department
Note: Do not sign your name. Your boss will not see this sheet. He or she will receive a
   summary of all responses from this department.
Consider your boss and how he or she performs compared to your expectations of him or
   her.
Does your boss:              Better than I would expect      Worse than I would expect
                                         10                              1
Know you? _____
Understand you? _____
Know what your needs are? _____

Write any comments here you wish

Back you? _____
Listen to you? _____
Talk to you? _____

Write any comments here you wish

Allow your input? _____
Ask for your ideas? _____
Use your ideas? _____

Write any comments here you wish

Remove any barriers in your way? _____
Have enough influence with his or her boss? _____
Have enough influence with other departments? _____

Write any comments here you wish

Talk down to you? _____
Treat you as a child? _____
Treat you as a subordinate? _____

Write any comments here you wish
```

Figure D-9. Inverse performance standards rating form.

Worker-Safety Analysis

Name _____ Date _____

Long-term analysis

   Biorhythmic information: dates to watch: _____
   LCU information: approximate units accumulated now: _____

Personality and value analysis

   Personality type _____
   Accident risk _____

	Key importance	No importance
Value of work		
Value of safety		

Current motivational analysis   Turn ons   Turn offs

   Peer group _____
   Me (boss relations) _____
   Company policy _____
   Self (personality) _____
   Climate _____
   Job-motivation factors _____
      Achievement _____
      Responsibility _____
      Advancement _____
      Growth _____
      Promotion _____
      Job _____
   Participation _____
   Involvement _____

Current job assignment   High   Low

   Pressure involved _____
   Worry or stress _____
   Information processing need _____
   Hazards faced _____
   Other _____

Force-field analysis

Pulls to safety
   ↑
――――――――――――――――――――――――――――――――――
   ↓
Pulls away from safety

Current assessment:

☐ OK   ☐ Discuss with worker ☐ Training ☐ Crisis intervention
☐ Crisis intervention   ☐ Contract   ☐ Behavior modification

Figure D-10. Worker safety analysis form.

Bibliography

Abercrombie, S. A. Enlarging the focus on motor fleet safety, *Professional Safety*, July 1983.

Albert, K. *How to Be Your Own Management Consultant.* New York: McGraw-Hill, 1978.

American Society of Safety Engineers. Scope and function of the professional safety position, undated.

Bigos, S. J., et al. A prospective study of work perceptions and psychosocial factors affecting the report of back injury, *Spine* 16:1–6, 1991.

Blake, R. P. *Industrial Safety.* Englewood Cliffs, NJ: Prentice-Hall, 1943.

Blanchard, K., and S. Johnson. *The One Minute Manager.* New York: Berkley Publishing Group, 1981.

Burke, A. The push to produce, *Industrial Safety and Hygiene News,* December 1994.

Business Management. How to conduct a post-mortem on a management consultant, 1965.

Business Week. The benefits of doing your own consulting, May 16, 1977.

Califano, J. Controlling health care costs, *National Safety News,* Chicago, January 1985.

Carey, H. Consultative supervision, *Nation's Business*, April 1937.

Chhokar, J. S., and J. A. Wallin. Improved safety through applied behavior analysis, *Journal of Safety Research,* 15(4):141–51.

Deming, W. Quality productivity and competitive position, unpublished report, MIT, 1982.

Fragala, G. A modern approach to injury recordkeeping, *Professional Safety*, January 1983.

Gleason, J., and D. Barnum. Effectiveness of OSHA penalties: myth or reality? Report from the Wisconsin Department of Industry, Labor, and Human Relations. In D. Peterson and V. Goodate, *Readings in Industrial Accident Prevention,* New York: McGraw-Hill, 1980.

Hales, T., et al. *Health Hazard Evaluation Report: HETA 89-299-2230, U.S. West Communications.* Cincinnati, OH: National Institute for Occupational Safety and Health, 1992.

Heinrich, H., D. Petersen, and N. Roos. *Industrial Accident Prevention,* 5th Ed. New York: McGraw-Hill, 1980.

Ishikawa, K. *Guide to Quality Control,* 2nd Ed. White Plains, NY: Quality Resources, 1986.

Johnson, W. G. *MORT—Management Oversight Risk Tree.* Washington, DC: United States Government Printing Office, 1973.

Kamp, J. Pre-employment personality testing for loss control, *Professional Safety,* pp. 38–40, June 1991.

Kamp, J. Worker psychology, safety management's next frontier, *Professional Safety,* May 1994.

Kennish, J. Violence in the workplace, *Professional Safety,* June 1995.

Komaki, J., K. D. Barwick, and L. R. Scott. A behavioural approach to occupational safety: pinpointing and reinforcing safe performance in a food manufacturing plant, *Journal of Applied Psychology,* 63(4):434–45, 1978.

Komaki, J., A. T. Heinzmann, and L. Lawson. Effect of training and feedback: component analysis of a behavioural safety program, *Journal of Applied Psychology,* 65(3):261–70, 1980.

Krause, T. R., J. H. Hidley, and S. J. Hodson. *The Behavior Based Safety Process.* New York: Van Nostrand Reinhold, 1990.

Likert, R. *The Human Organization.* New York: McGraw-Hill, 1967.

Lippett, G. *Organizational Renewal: Achieving Viability in a Changing World.* Englewood Cliffs, NJ: Prentice-Hall, 1973.

Mager, R. *New Patterns of Management.* New York: McGraw-Hill, 1961.

McBride, G., and P. Westfall. *Shiftwork Safety and Performance.* National Safety Council, Chicago, 1992.

National Safety Council. *Accident Facts.* Chicago.

National Safety Council. *Motor Fleet Safety Manual.* Chicago, 1966.

Newport, D. A review of training fundamentals, *Training and Development Journal,* 22(10), 1968.

Peters, G. Why only a fool relies on safety standards, *Professional Safety,* January 1966.

Peters, T., and R. Waterman. *In Search of Excellence.* New York: Harper & Row, 1982.

Petersen, D. *The OSHA Compliance Manual.* New York: McGraw-Hill, 1975.

Petersen, D. *Human Error Reduction and Safety Management.* Goshen, NY: Aloray, 1980.

Petersen, D. *Safe Behavior Reinforcement.* Goshen, NY: Aloray, 1989.
Petersen, D. *Techniques of Safety Management,* 3rd Ed., Goshen, NY: Aloray, 1989.
Petersen, D. *Statistical Safety Control Manual.* Tempe, AZ: DPA, 1990.
Pope, W. C. In case of accidents, call the computer, *Selected Readings in Safety.* Macon, GA: Academy Press, 1973.
Pope, W. C., and T. J. Cresswell. Safety programs management, *Journal of the ASSE,* August 1965.
Rhoton, W. W. A procedure to improve compliance with coal mine safety regulations, *Journal of Organizational Behaviour Management,* 2(4):243–49, 1980.
Rinefort, F. A new look at occupational safety, *Professional Safety,* September 1977.
Siegel, J. *Managing with Statistical Methods.* Society of Automotive Engineers, 1982.
Simonds, R. H., and Y. Safai-Sahrai. Factors apparently affecting injury frequency in eleven matched pairs of companies, *Journal of Safety Research,* September 1977.
Smith, M. J., W. K. Anger, and S. S. Uslan. Behavioural modification applied to occupational safety, *Journal of Safety Research,* 10(2):87–8, 1978.
Sulzer-Azaroff, B. and C. M. de Santamaria. Industrial safety hazard reduction through performance feedback, *Journal of Applied Behaviour Analysis,* 13(2):287–95, 1980.
Sulzer-Azaroff, B. and D. Fellner. Searching for performance targets in the behavioral analysis of occupational health and safety: an assessment strategy, *Journal of Organizational Behaviour Management,* 6(2):53–65, 1984.
Walter, M. *The Deming Management Method.* New York: Dodd, Mead, 1986.
Weaver, D. A. Symptoms of operational error, *Professional Safety,* October 1971.
Zebroski, E. Lessons learned from man-made catastrophes. In *Risk Management,* by R. Knief et al., Washington, DC: Hemisphere Publishing, 1991.

Index

Accident investigation, 253–254
 analysis tools, 50–54
 procedures, 47–54, *52*
Accidents
 causation, *51*, 185–187
 management and, 187
 severe, 188–190
Accident statistics
 validity of, 4, 15–16, 33
Accountability systems, 94–95, 190–191, 247–252
 MENU, 251–252
 Safety by Objectives, 249–251
 SCRAPE, 247–249
Albert, K., 228–229
Alcohol, 117–118, 120
American Society of Safety Engineers (ASSE), 4, 5, 7
Analysis of safety system effectiveness
 by workers, 33–41
 research, 25–29
 survey results, 171–184
 three-step process, 3
 traditional methods, 15–24
Anxiety, 79, 86. *See also* Aversives
Areas to analyze, 25–29, 43–45, 135–136
Attitudes toward safety, 75, 77–78
Audits, 4, 5, 16, 128, 205
Aversives, 78–80
Awareness programs, 121–126

Barnum, D., 131
Behavior, 121–125, 191–192
 improvement of worker, 115–126
 modification, 74–75
 observations, 30
 patterns, 86
 sampling, 33, 40–41, 239–246. *See also* Safety sampling
 tracking, 152, *153*
Benchmarking, 29–31
 communicating, 29, 30
 safety leadership, 29–30
 safety performance, 29, 30
 training for safety, 29, 30–31
 best practices, 29–31
Bioecological stressors, 85–86
Biofeedback, 87
Biorhythms, 85
Blanchard, K., 153
Boredom, 80. *See also* Aversives
Brereton, P., 91
Burke, A., 151

Carey, H., 55
Catastrophe control, 136–139
Catastrophe vs. attribute matrix, *138*
Centers for Disease Control, 83–84
Challenge of Change, The, 33
Champion for Safety, 38

267

Change, implementing, 193–208
Climate, 65–66. *See also* Culture
  analysis form, *260*
  and learning, 105
Communications, 29, 30, 108, 110–114, 115–117
  checklist, 117
  management variables, 110–112
  message variables, 112
  model, *110*
  receiver variables, 112, 114
Cognitive appraisal, 87
Conditions, positive and aversive, 78–81
Consequences, positive and aversive, 79
Constancy of purpose, 200. *See also* Deming Management Method
Consultants
  analyzing performance, 223–228
  choosing, 223
  government, 211, 215–220
  internal, 228–229
  insurance, 210, 213–214
  private, 211–212, 220–223
Contacts, safety. *See* Safety contacts
Control charts, 95, 206, *207*
Core Media, Inc., 33
Cost containment, 6, 166–169
Cresswell, T., 8, 191
Creative goals. *See* Goals for safety performance
Credibility. *See* Management credibility
Critical Incident Technique (CIT), 50
Culture, 11–14, 49, 65–68, 192. *See also* Climate
  and safety, 66–67
  building, 65
Cumulative Trauma Disorders (CTDs), 160–166, 189

Data
  interpreting, 171–184
  uses of, 207
Dawson, L.H., 19
Decision to err, 49, 191
Deming, W.E., 154, 199, 205, 207
Deming Management Method, 199–201
Deprivation, 85
Design, 49, 158–159, 201. *See also* Ergonomics
Diet, 87

Discipline, 30, 59–64, *63*
Documentation, 131
Drugs, 117–120

Embarrassment, 80. *See also* Aversives
Employee(s)
  involvement, 14, 55–57, *58,* 154–156, 198–200, 208
  new, 105–108
  participation, 55–57
  problem, 119–120
  role, 10–11
  screening, 119
  selection of, 105, 107–108
  training, 101–105
Employee Assistance Programs (EAPs), 87, 120–121, 143, 165
Ergonomics, 49, 159–166, 201
  Job analysis chart, *163*
Exercise, 87

Fault trees, 50
Fear, 79. *See also* Aversives
Feedback, and learning, 104
Fight or flight reaction, 85
First line supervisor, 10. *See also* Management
Fishbone diagrams, 50, 53–54, 203–205, *204*
Flow charts, 95, 205, *205*
Frequency rates, 15–16
Frustration, 79–80, 85. *See also* Aversives.

Gimmicks, 124–125
Gleason, J., 131
Goals for safety performance, 95–100
  creative, 96, 97
  criteria, 98–100
  personal, 96, 97
  project, 96, 97
  results, 98
  routine, 96–97
Goal setting, 86

Hawthorne effect, 15, 154, 199, 200
Hazard control, 130–132
Hazard correction, 130–134
  matrix, *132*

Index

Hazard hunt, 259
Heinrich, H., 8, 17, 127, 152, 155, 185, 188
Hersey, P., 153
High performance organizations, 152–157
Histograms, 202, *202*
Human error, 47–50
Human factor, 49
Humiliation, 80. *See also* Aversives

Illnesses, stress-related, 81, 82–85. *See also* Stress
Incident Recall Technique (IRT), 258
Individual strategies, 87. *See also* Stressors
Insomnia, 151
Inspections, 127–129, 254, 255
Insurance companies, 210, 213–214. *See also* Consultants
Interviews, 33, 36–40
Inverse performance standards, 260–261

Job analysis, 90
Job enrichment, 86
Job Safety Analysis (JSA), 12, 57, 66, 67, 254, *256*, 261
Job Safety Observation (JSO), 255, *256*, 257
Johnson, W.G., 17–19, *18*
Juran control cycle, 17, *18*

Kamp, J., 162, 163
Kennish, J.W., 139–143
Kepner-Tregoe sequence, 17, *18*
Key person, 10, 187

Likert, R., 65–66
Lippitt, G., 220–222

Mager, R., 77–78
Management
  commitment, 12–13
  communications, 110–112
  credibility, 68–70, *71*
  involvement, 14
  model, 3
  performance, 89
  policy, 48, 69, 145
  responsibility, 101
  role, 10, 70, 72
Management by objectives (MBO), 17, *18,* 250
Management Oversight Risk Tree (MORT), 17–24
Measurement (of safety performance), 93, 231–238
Message variables, 112. *See also* Communications.
MENU, 251, *251–252*
Middle management, 13, 72, 89. *See also* Management
Modeling, 81. *See also* Behavior
Motivation, 103–104. *See also* Training
Multiple causation, 8–10

National Commission on Sleep Disorders, 151
National Institute for Occupational Safety and Health (NIOSH), 140, 164
Nertney hazard analysis wheel, *19*
Newport, D., 102
Noise, 85
Nutrition, 85

Obstacles, 12–13. *See also* Safety
Occupational Health and Safety Act (OSHA), 6, 16, 131
  consultants, 210, 211, 215–216
  guidelines, 59, 66–67
One-on-ones, 257
Operating procedures, 57, 59, *60, 61*
Operational errors, 9
Organizational analysis, 90
Organizational strategies, 86. *See also* Stressors
Orientation, 92
Outside help, 209–228. *See also* Consultants
  sources, 210–223
  when to seek, 209–210
Overload, 48–49, 85, 150–151, 191
Overtime, 150–151
Ownership, 29–30. *See also* Benchmarking

Pain, 79. *See also* Aversives
Pareto charts, 95, 202–203, *203*

Participation, 55–57. *See also* Employees
Perception surveys, 14, 33–36, 74, 174–175
 differences, *183*
 interpretation, 176–184
 results, 193–198
 vs. interviewing, 40
Performance models, 50, 90–92
Personality stressors, 86
Peters, G., 216–220
Peters, T., 65
Physical conditions, 127
Physical discomfort, 80. *See also* Aversives
Pope, W., 8, 191
Positive reinforcement, 74–75
Positives, 78–81
Preemployment screening, 119
Problem employees, identifying, 119–120. *See also* Employees
Psychosocial stressors, 85
Purchasing, 158

Recognition, 74–75
Reengineering, 154
Regulatory compliance, 16–17
Reinforcement, 104. *See also* Training
Relaxation, 85, 87
Restructuring, 87
Rewards, 71, 92, 93–94
Rinefort, F., 25–29
Rightsizing, 154
Roles, in safety, 70, 71–73, 191
Role specification, 86
Run charts, 206

Safe behavior reinforcement, 80–81
Safety
 attitudes toward, 75–78
 contacts, 115–117
 corporate obligations, 201
 criteria for effectiveness, 192
 function of, 7–10, 187–188
 leadership, 29–30
 obstacles, 12–13
 research on effectiveness, 25–29
 responsibilities, 7, 69
 support for, 71–73
 ten basic principles, 188–192

Safety by Objectives, (SBO), 249–251
Safety champion (among managers), 29
Safety culture, 12, 14, 65–66. *See also* Climate, Culture
 obstacles, 12–13
Safety management principles, 7–10
Safety performance, 30
 analysis of, 17
 goals for, 95–100
 recognition for, 74–75
Safety professional's role, 4–7, 72, 187–188, 191
 OSHA responsibilities, 6–7
 safety responsibilities, 7
 scope and functions, 4–6
Safety programs, 4, 17
 and culture, 12
 tools, 66–67
Safety sampling, 40–41, 239–246, 257–258. *See also* Behavior sampling
Safety systems
 areas to analyze, 25–29, 43–45
 assessing, 3, 4, 15–17
 effectiveness, 15, 16
 management approaches, 17–24
Scatter diagrams, 95, 205, *206*
SCRAPE, 247–249
Self-perception, 86
Shafai-Sahrai, Y., 25
Shift work, 150–151
Simonds, R., 25
Sleep disorders, 151
Society of Automotive Engineers, 218–219
Standard operating procedures (SOPs), 57, 69
Standard procedure instructions (SPIs), 57
Statistical Process Control (SPC), 53, 95, 154, 198–208
 control tools, 206
 input data, 207–208
 problem-solving tools, 202–205
Statistics, accident. *See* Accident statistics
Stress, 81–88
 control programs, 86–88
 diseases, 84
 natural responses to, *84,* 85–86
 related-illnesses, 81, 82–85
Stressors, 85–87
 bioecological, 85
 personality, 86
 psychosocial, 85

Index

Supervision, quality of, 92
Supervisory training, 89–92
Supervisory performance, 71–72, 253
   effectiveness of, 94–95
   model, 93–94
Supervisory role, 72–73
   nontraditional tasks, 72, 257–262
   traditional tasks, 72, 155, 253–257
Survey results, interpretation of, 176-184
   clusters of categories, 178–180
System failure, 48

Taylor, F., 152
Teams, 154–156
Technique of Operations Review (TOR), 50, 258–259
Techniques of Safety Management, 7
Three-step process, 3. *See also* Analysis of safety system effectiveness
Total involvement concept, 154–155
Total Quality Management (TQM), 53, 154, 198–200
Training, 29, 30–31, 86
   employee, 101–105
   evaluation of, 105
   supervisory, 89–92
Traps left for worker, 49, 191, 202

Upper management, 13, 14, 25, 69, 70, 71, 73. *See also* Management
Upward communication, 116. *See also* Communications

Vehicle accident control, 144–145
   driver selection, 145–147
   driver training, 147–148
   policy, 145
   records, 149–150
   vehicle maintenance, 148–149
Violence in the workplace, 139–143

Weaver, D.A., 8, *9*
Wellness programs, 86
Work scheduling, 86
Worker analysis, 90–91, 261, *262*
Worker's compensation, 83, 164–166